www.powys
D0707997
Llyfrgell y Trallwng
Welshpool Library
Ffôn/Tel: 01938 553001
PRESTEIGNE
LIBRARY
01544 260552
21 SEP

Powys
37218 00405686 2
Cancelled

natural Dog Care

natural Dog Care

Christopher Day

hamlyn

An Hachette UK Company
www.hachette.co.uk

First published in Great Britain in 2011 by
Hamlyn, a division of Octopus Publishing Group Ltd
Endeavour House, 189 Shaftesbury Avenue
London WC2H 8JY
www.octopusbooks.co.uk

ISBN 978-0-600-62116-4

A CIP catalogue record for this book is available from the British Library

Printed and bound in China

10 9 8 7 6 5 4 3 2 1

Note

Unless the information is specific to males or females, throughout this book dogs are referred to as 'he' and the information and advice are applicable to both sexes.

The advice in this book is provided as general information only. It is not necessarily specific to any individual case and is not a substitute for the guidance and advice provided by a licensed veterinary practitioner consulted in any particular situation. Always seek professional advice in any cases of disease or illness. Octopus Publishing Group accepts no liability or responsibility for any consequences resulting from the use of or reliance upon the information contained herein.

No dogs were harmed in the making of this book.

CONTENTS

INTRODUCTION

Dogs have earned a place in our homes and hearts, almost as if they are part of the human family. They are treasured companions, friends and pets, who, in their turn, are loyal, loving and devoted members of the household.

When we take on a dog, whether it is a puppy or an adult, we want everything that is best for him, in order to ensure his welfare, happiness and fulfilment. We want him to lead as active and as healthy a life as possible, for as long as possible.

We also take on a responsibility and a duty of care for a pet. Of course your veterinarian is the first port of call, should anything go wrong with your dog's health, but it is more constructive to think of positive health and prevention of disease – through healthy lifestyle and environment – as the best insurance against illness.

While modern Western medicine has much to offer in the case of disease, ill health is still prevalent. In the USA several million dogs are diagnosed with cancer each year and one in three dogs dies of cancer, the biggest killer of dogs over two years old. Arthritis, heart disease, autoimmune disease, allergies, skin problems, colitis, diabetes and other chronic

▼ *Our dogs should lead active and healthy lives.*

diseases blight the lives of countless more dogs. All is not well, despite the heavy investment in veterinary medical research and the development of modern medicines.

No wonder more attention has been focused lately on natural and holistic lifestyle methods, both for our family and for our animals. There is a great deal of logic behind the notion that, if our dogs were to eat more healthily, drink good water, exercise well and regularly, enjoy a lifestyle that is as free from stress as possible while enabling behavioural needs to be fulfilled, then the health of our dogs would be better and their resistance to disease enhanced. If we also restrict their chemical exposure and use gentle, natural medicine techniques (should their health begin to waver) in order to stimulate the body to regain health, then we avoid the risk of adverse drug effects, unless drugs become absolutely necessary to maintain welfare or to preserve life itself.

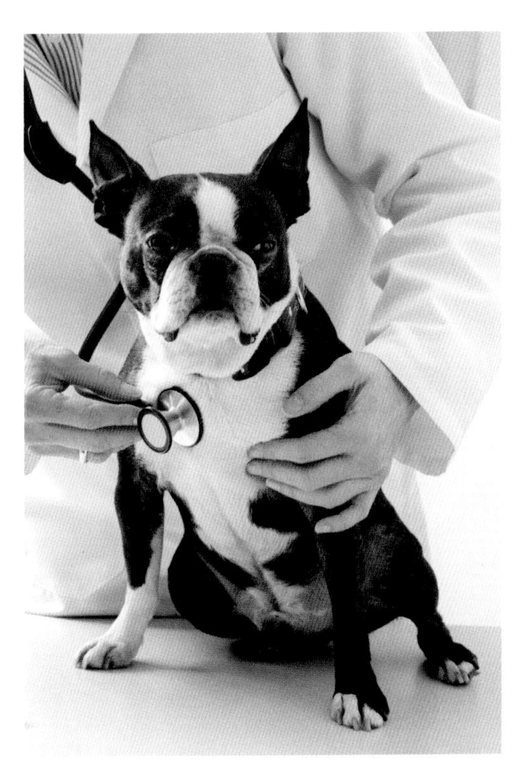

▲ *Your veterinarian can monitor your dog's health and diagnose disease.*

THE AIM OF THIS BOOK

This guide is designed to help the caring dog owner to encourage positive health and establish healthy routines, in order to improve the health and welfare of the family dog. It has often been said that prevention is better than cure. Certainly it is preferable to order our daily routines so that we do as much as possible to prevent illness and avoid the need for veterinary intervention. However, this book is not designed to act as a substitute for proper veterinary care, should disease or illness occur. A visit to your vet for diagnosis and assessment is recommended whenever illness affects your dog. If you are lucky enough to have a vet in the neighbourhood who is trained both in modern conventional medicine and in holistic and natural medicine, so much the better, because all available medical options can then be considered and an objective and balanced medical programme can be devised.

This book is a convenient and compact introduction to the options available, and does not pretend to be a complete treatise – a massive and unwieldy volume would be required for that purpose. However, it should act as a guide to the health benefits that can accrue from a natural lifestyle, and as a pointer and aide-memoire in many health and injury situations that might arise in the home, and which are amenable to home care.

ABOUT DOGS

Dogs are immensely social creatures, who have descended from a wild ancestor, almost certainly a wolf-type animal. It is thought that they started to become domesticated somewhere between thirty thousand and fifteen thousand years ago.

▲ *All domestic dogs have descended from a wolf-type ancestor.*

The process of domestication is likely to have been a long one, possibly triggered by dogs scavenging around human 'kills' and the detritus of human dwellings, thereby becoming a common sight to human eyes. This process was probably fuelled by human beings' need to hunt wild animals, to protect dwellings and (eventually) to satisfy a yearning for companionship, all of which requirements a dog could fill.

Despite the massive diversity of modern dog breeds, all belong to the one species (*Canis lupus familiaris*). The dog's lupine origins explain much of what we need to know about it. Dogs are social animals, with a tendency to live in packs, with a recognized 'top dog'. Their interactions with others of the same species are governed by their primeval inherited instincts of pack behaviour. They tend to hunt in packs, with a complex range of cooperation behaviours, which make catching prey more certain. They are able to catch, hold, kill and dismember prey carcasses, often much larger than themselves (it is this trait that makes dogs dangerous for farm animals, if not properly controlled and trained, particularly if two or more dogs roam farmed land unsupervised). They are also capable of chewing bones. They are meat-eating predators and scavengers with an omnivorous tendency (eating fruits, roots and other vegetable matter, in addition to meat and carrion).

DOGS IN DOMESTICITY

In the domesticated context, dogs become devoted to the human 'pack leader'. Their general tendency is to try to please this perceived leader, whenever possible. They derive comfort from the pack leader's rewarding behaviour and from the presence of a human companion. They tend to protect the pack's lair – in this case, the human home. Humankind has, over the millennia, exploited various physical and behavioural attributes of the canine species and enhanced them by selective breeding. The dog's powerful vision, sensitive hearing and smell, hunting instinct, herding instinct and retrieving, guarding and protecting behaviours have all been targets of selective breeding (man-made and man-accelerated 'natural selection'), arriving at many and varied breeds suited to certain purposes in the service of humans. Dogs hunt with us, detect game, control pest rodents, guard and help manage flocks and herds, seek out explosives and drugs, go to war and serve as our companions and pets.

When breeding, bitches give birth to a number of temporarily sightless, very defenceless and dependent young (called a litter). The usual number of mammary glands and teats is ten. In addition to suckling their young, dogs regurgitate food for the puppies, until the pups become less dependent. A bitch tends to come into 'season' about twice yearly (an individual can vary widely from this, but normally follows her own regular pattern) and produces blood in the week or ten days preceding ovulation. At the end of this bleeding phase, she is usually receptive to the male.

▲ *Domestic dogs come in a great variety of shapes and sizes, with different perceived roles for human interaction.*

SOME POPULAR BREEDS OF DOG

Bedlington Terrier	Hungarian Vizsla	Rottweiler
Border Collie	Jack Russell Terrier	Rough Collie
Border Terrier	Labrador	Schnauzer
Borzoi	Lhasa Apso	Sharpei
Cocker Spaniel	Miniature Poodle	Shetland Collie
Dachshund	Miniature Schnauzer	Springer Spaniel
Deerhound	Münsterländer	Staffordshire Bull Terrier
Dobermann Pinscher	Norfolk Terrier	Tibetan Terrier
English Bull Terrier	Pekinese	Toy Poodle
Golden Retriever	Poodle	Whippet
Great Dane	Pug	Wolfhound
Greyhound	Rhodesian Ridgeback	

ABOUT PUPPIES

It is during puppyhood that so many physical, behavioural and health patterns become established, hence the extreme importance of careful holistic rearing. Puppies are born after about nine weeks of gestation.

After about 12 days the eyes open to give the pup limited vision. His visual needs at this stage are restricted to the nest area. Sight develops progressively, as does independence. The dog's skeleton is very soft and pliable at first, being composed mainly of cartilage, until ossification gradually turns cartilage to bone during the puppy's subsequent growth and development.

Play between littermates starts to unfold the instinctive behaviour of the growing dog. The mother will also teach certain behaviours, during play and interaction. This phase is vital to the eventual social development of the adult. Hence a puppy should not be removed from his mother before eight weeks of age.

▲ *Ball play can overstrain a very young puppy.*

FEEDING AND CHEWING

The pup is able to deal with regurgitated meat or meat scrapings from an early age. The 28 puppy teeth erupt from the third week of life. These will eventually irritate the mother during nursing, so gradual weaning will commence. However, it is not until the full complement of (normally) 42 adult teeth develops, between four to seven months of age, that the pup can deal really effectively with chunks of raw meat and bones. If a natural diet is planned, then it is important that the young pup is given bones (see page 20) and meat during his development, in order for him to learn the necessary skills and habits for safe application of such a diet.

A young puppy usually wants to chew, especially during periods of teething. This need can be satisfied by providing bones and chunks of meat, thus saving damage to furniture, fittings and clothing and preventing the subsequent friction between human and dog that may occur from such damage. Furthermore, exposed electric cables seem to be endlessly fascinating to a teething puppy, with risk of serious injury or death.

▲ *A bitch may have many puppies.*

It is important not to overfeed or overexercise a young puppy, especially in the case of large breeds, so that skeletal growth and development can occur in a healthy and timely way. If growth is too rapid or exercise too strenuous too early, then damage to the young skeleton can occur, leading to problems in adult or later life.

PLAYING WITH AND HOLDING A PUPPY

When playing with a young puppy, remember that the milk teeth are very sharp indeed. If there are young children in the household, they must be supervised when playing with the pup, so that both learn together how to play sociably and safely.

If a pup is to be picked up, it is important to exercise great care not to drop him; he may wriggle enthusiastically and boisterously. While the pup is discovering his home environment, be very careful with doors in the house and in the car to avoid injury.

PUPPY CHECKS

- It is valuable to get a health check done by your local vet, who is trained to spot early signs of disease and developmental or congenital issues.
- In a male puppy, check that both testes have descended, and seek advice if not.
- At about six months of age, the puppy's teeth should be checked for proper deciduation – the process by which milk teeth fall out and are replaced by adult teeth.

See also *About Dogs (8–9), Natural Training (16–17), Vaccination and Alternatives (34–37), Worms (40–41)*

EXERCISE AND PLAY

Exercise and play are essential to the growing puppy, to continue his education and socialization, to encourage a healthy heart and circulation, a robust skeletal structure and muscular health. In fact, all life functions depend upon the challenge of beneficial exercise, for healthy development and maintenance.

▲ *Lead-walking is important in busy public places.*

Play is important for the puppy in order to learn about action and reaction, to bond with the human family and to maintain a healthy outlook on life. However, some individuals will be willing to play until they drop, with serious potential health implications to the musculoskeletal system and the heart. Too much boisterous play in hot weather is especially dangerous. Ball play can also be fast and 'extreme', with a serious risk of injury and exhaustion. Even an adult dog, if ball-obsessed, will not learn the limits. Terriers, Collies, German Shepherds and Labradors are especially known for this.

It is important to teach children to know the limits, too. Playing with an energetic and fun-loving pup can be so absorbing that the danger signs are not heeded. Supervision is essential until a safe routine has been established.

SOME SOUND GUIDELINES

Never allow a dog to play 'fetch' with a stick. This is a common cause of severe injury or death from impalement, for the timing and landing of a stick can never be fully controlled. A dog can rush onto the

▲ *Never throw sticks for a dog as this is a common cause of injury.*

end of a stick, with gaping mouth, just as the stick strikes the ground. The momentum of the dog can cause a deep penetration injury, into throat, neck or chest.

Apart from periods of play, regular exercise – two or three times a day – is fundamental to a dog's well-being. Depending upon the breed, upon the individual dog and upon the mobility and circumstances of the owner, the distance and time can vary. For the sake of public relations for the canine world, you should remember to keep your dog on a lead in specified areas, pick up faeces and not allow your dog to be a nuisance to other people or other dogs.

In the countryside, it is worth being wary of livestock. Dogs, particularly if they are with a canine companion, are very attracted to running sheep. This can lead to sheep-mauling or killing, even in an otherwise well-behaved dog. Dogs may also chase other livestock, or can even be chased menacingly by cattle and horses, with attendant dangers to a human companion. It is also worth remembering that a dog can impulsively chase a cat, even across a busy road, if it is not on a lead.

On a beach it is important to respect other users and to clear up after your dog. There may be special laws or restrictions for dogs on certain beaches. If you are near the sea, a river in spate or frozen water, be careful not to allow the dog to fall into the water. Many owners have been killed in an attempt to rescue a dog, with the dog itself often surviving.

TRAINING AND RESTING TIME

The pleasure of exercising a dog will be greatly enhanced by good training. When exercising a dog on a lead, a harness may be preferable to a collar, which can pull harshly on the dog's neck. Halters and similar restraints may help to control a pulling dog, but they can snatch dangerously at his neck.

All dogs require periods of quietude and rest during the day. At home, they should have a designated area where they feel secure, and children should be taught to respect their resting time.

See also *Natural Training (16–17)*

MASSAGE

Massage is a form of physical or tactile therapy that can bring multiple benefits. It relaxes the muscles; it promotes the removal of toxins and by-products from muscles and soft tissues by improving blood flow and lymphatic drainage. This in turn reduces the tendency to stiffness and also brings incidental benefit to the joints.

Massage is best performed on a dog that has settled after exercise, and should be done when you are calm, too. It should be performed on soft tissues only – massaging over the bones or the spine can be unpleasant for the dog and may cause trauma. It should include all muscle areas, including the muscles of the head, face and abdomen. It should generally be done in the direction of the muscle fibres (in the direction of the pull of the muscles), wherever possible.

Routine massage, when there is no particular problem to address, should be a 'whole body' treatment, sometimes called effleurage. This is basically a gentle stroking action over all the muscles of the body, without dragging the skin. If regions of tightness or spasm are found, then deeper and more directed massage can be given. Cords and knots can be felt in affected muscles and can be released by gentle but firm pressure and movement. The pressure and duration of the massage should be guided by the dog's responses. Where there are regions of tightness and spasm, be alert to the possibility that this could indicate spinal misalignment, which should be treated by chiropractic manipulation.

1 Both you and your dog should be relaxed, and he will soon understand when you are ready to offer massage. Apply a gentle and even pressure in flowing movements (*effleurage* is the French for 'skimming') on each side in turn, from behind the ear and down the front of the shoulder to the front of the chest. Repeat three times.

Dogs will soon settle into a routine, if massage is performed regularly. They usually find it therapeutic, as does the masseur or masseuse.

While performing the overall massage, take the opportunity to check for anything untoward, which may have gone unnoticed at a purely visual inspection.

See also *Chiropractic and Osteopathy (58–59), Other Natural Therapies and Treatments (60–61)*

FEEDBACK FROM YOUR DOG

Massage is a very rewarding pastime, for dog and owner alike. It is a hands-on therapy that anyone can learn. A dog will clearly demonstrate when it enjoys or needs a massage, and will find a way to indicate whether more or less pressure is required and when the masseur/masseuse should move to a different site or cease the session.

2 Massage from the top of each shoulder to the elbow, then down the front and outside of each front leg to the paw, then up the back and inside of each front leg. Repeat each stage three times.

3 With your dog standing, massage from the top of each shoulder along each side of the back to the tail, avoiding the spine. Then massage each side, from hip to stifle (the dog's 'knee'), and then down the front and outside of each back leg to the paw, then up the back and inside of each back leg. Repeat each stage three times.

NATURAL TRAINING

It is so much easier and more fun to live with a well-behaved dog. Depending upon the individual, this may require a great deal of input from the start or may simply be a matter of encouraging those behaviours that you want and discouraging those you do not. So many dogs are so eager to please that they almost train themselves.

▲ *A well-behaved dog is a joy and makes for safety outside the home environment.*

Consistency and self-discipline are the first prerequisites for successful training. Commands should always be repeated simply and clearly. Each time an infringement or compliance occurs, reward or discouragement should be given accordingly.

EXAMINING THE DOG

At the very least, a dog should be willing to submit to careful examination and handling, including the feet, ears, teeth, mouth and under the tail. If no-go areas are established, then (should an injury or illness occur) it may be a source of great trauma to examine and deal with those areas. From early puppyhood it is therefore a good idea to carry out mock examinations and handling of all parts of the body. It is also worthwhile holding the puppy, carefully and gently, in many different positions. Strong-willed puppies may object to some part of this routine, but this resistance must be overcome with firmness, love and persistence, to ensure that the puppy accepts the authority of the 'pack leader'. A puppy with no fear of being restrained, turned onto his back and having his mouth opened will be much happier at the veterinary clinic, should the need arise.

REMOVING FOOD AND TOYS

Another vital area of potential conflict that should be averted is that of food and bones. From an early age the pup should willingly allow you to take food away, remove a bone, handle him while he's feeding and offer other indicators of acceptance of your natural authority. Allowing a puppy to guard food or toys jealously will only set up a store of trouble for later. Also, should a dog pick up something potentially dangerous, removing it successfully or ensuring that the dog leaves it should occur without delay and may save his life.

VERBAL COMMANDS

Understanding the word 'no' to mean stop whatever he is doing may also be a potential life-saver and will certainly make for a better relationship throughout the dog's life. The word should always be used firmly and clearly, and your dog should be taught to respond promptly.

Simple commands such as 'sit', 'stay', 'down', 'over' and 'bed' are useful and can be vital in stressful situations. Other helpful commands are 'here' and 'heel'. These can be taught to a dog in a very short time, with frequent repetition and reward, and by association (by using the commands whenever a dog performs the relevant procedure of his own volition, during the day).

HOUSE-TRAINING AND LEAD-TRAINING

House-training is rarely a problem, but again consistency and sensitivity to the dog's needs are essential. Reward or discouragement can be given by body language and by tone of voice, expressing either joy or disapproval.

For most households, lead-training is vital and should be started from an early age. Kindness, consistency and making it fun should speed up the process and make it enjoyable for both parties. Collars or harnesses should be comfortable and well fitted, and a 'choker' or halter should not become necessary if training takes place early enough, never permitting unsuitable behaviour patterns on the lead to develop.

One way of rewarding is by giving wholesome, tasty treats but beware of overdependence on these.

GROOMING, TOENAILS, TEETH AND EARS

A wild dog grooms himself and maintains his claws, skin, teeth and coat in good condition. However, certain inherited characteristics in dogs, or an unnatural diet or unsuitable exercise regimen, can lead to a need for special attention.

GROOMING

Dogs with long coats require regular combing and brushing. Tangling or matting of the coat can lead to skin disease. The dog should enjoy this attention from the outset. If this proves difficult at home, seek help from a professional dog groomer. Some dogs benefit from clipping in early summer, to prevent overheating.

▲ *Regular grooming is vital for dogs with long coats.*

TOENAILS

A dog who has plenty of exercise, including on abrasive or hard surfaces, will usually keep his toenails healthy. But many dogs require some help in maintaining toenail length and condition. This can be a result of conformation (if the foot is flat, the toenails will grow longer before wearing on a surface) or because the exercise regimen is not ideal. A vet is usually very willing to help with clipping toenails, but it is also something that can be done at home.

A good-quality set of toenail clippers is required. It is important that the dog should not be hurt by the procedure. Many dogs acquire extreme (and quite understandable) fear of the procedure, having been cut too close to the quick or having had the quick pinched by an incorrect procedure. In white claws, it is usually easy to see the quick. A cut should not be made closer than about 3 mm (1⁄10 inch) from the tip of the quick. In dark claws, great care is needed, and veterinary help may be the wiser option. The cut should never be made from side to side (or painful pinching can result, even without drawing blood), but always from the dorsal to ventral (from the ground side to the upper side of the nail). It is important to check all dewclaws too.

TEETH

A natural-type diet of raw meat with an occasional raw knuckle bone (see page 20) to chew may prevent problems with teeth and gums (a marrow bone may be too rich and too hard for safety). Despite the simplicity of this preventive measure, many dogs need dental attention because of starchy, dry or sloppy foods. The teeth need to be subjected to the stresses and strains of chewing raw meat and bones to maintain healthy function. Should tartar begin to develop, it needs to be cracked or scraped away before it causes gum recession. Minor scaling can usually be easily be removed by the thumbnail. Plaque can be cleaned off with a toothbrush. In cases where there is already bad oral health, it may be necessary to ask a vet for help, in order to make a fresh start.

EARS

In most dogs, the ears require no attention. They should be checked from time to time, nonetheless. Signs of trouble may be scratching or shaking the ears, with the affected ear being held low. If the dog suffers from a skin problem, the ears may also be affected, resulting in inflammation, excess wax production or even pathological discharge. Veterinary help may be needed, but it may be possible to obtain sufficient relief by the use of aromatherapy. In dogs with generalized skin and ear problems, a homeopathic constitutional prescription should be sought from a holistic vet. Dogs with hairy, long ears (Spaniels, for example) may be prone to picking up grass seeds in their ears, especially during summer or autumn. This is an extremely painful condition and requires immediate veterinary help. Clipping such ears during early summer may prevent the problem.

▲ *The quick shows as a pink cone within the claw.*

See also *Diet (20–21), Aromatherapy (54–55)*

DIET

A dog's closest living equivalent is the wolf or wild dog. Keeping the diet of a wolf or wild dog in mind is a good guide to how your own dog should be fed, for maximum health and longevity.

Wolves and wild dogs are by nature carnivorous, omnivorous and scavenging. They catch and devour wild living prey, consume dead animals (carrion), dig up roots, eat fruit and berries, eat herbivore dung and the ingesta (gut contents) of their herbivore prey. When eating carrion or prey, they will consume hair, skin and bone, in addition to cartilage, sinews and ingesta. This ancestral diet is very informative. It is logical to assume that, if we stray too far from this with our own pets, ill health and poor growth and development are likely to ensue.

SUITABLE AND UNSUITABLE FOODS

- Wholesome, fresh, raw meat (not pig) is better than cooked. Giving raw bones is also possible. However, caution is advised when giving an adult dog bones for the first time, since his instinct may not be very strong, resulting in some risk. The mineral benefits of bones can still be obtained by grinding raw bones *thoroughly* for dogs that are unsafe with whole bones.
- Dogs require vegetable material, but in processed form (they cannot chew and, in the wild, their vegetation intake is partly digested). Vegetable material can be liquidized or blended, boiled or steamed. If a juicer is used, feed the pulp (fibre) too. Carrots should be organically grown.
- Dogs can eat fruit, but research suggests that grapes (and currants, raisins or sultanas) may be toxic to some dogs.
- Animal fat (raw), vegetable oils (not solvent-extracted), additive-free fish oils, additive-free cod-liver oil and essential fatty acids (omega-3, 6 and 9) are suitable supplements for dogs.
- Grain starch contains little nutrition and is not required. In some cases, it can be harmful. It is not part of the diet of a wild wolf, although a very 'high-energy' dog may be able to use starch satisfactorily.
- Cow's milk and dairy products (butter, cheese and cream), especially if pasteurized, should be avoided wherever possible. Yoghurt and cottage cheese may be acceptable, as a result of the process involved in their preparation. Experience suggests that goat's milk and its products are well tolerated.

CAUTION

Whether feeding suitable raw bones represents a health risk is a matter of opinion and depends on the individual dog. If in doubt, seek the advice of a holistic vet.

- Avoid both salt and sugar.
- Chocolate can be toxic to dogs.
- Water should be filtered or obtained from a spring or well. Softened water is usually unsafe. When buying water or when travelling, glass bottles are preferred to plastic, which can leach toxins into the water.

▼ *Pottery or ceramic bowls, dishes and plates are recommended.*

FOOD AND DRINK CONTAINERS

The receptacle in which in the food or water is offered can be very important. Ceramic, china, terracotta or other pottery-type bowls, plates or dishes are recommended because they do not leach undesirable or toxic material into the diet. Enamelled metal bowls may be acceptable. Plastic or stainless-steel bowls are not recommended. Plastic is toxic, the colourants used are toxic, and stainless steel may leach nickel (allergenic) and other metals into the contents. If special dishes are required (for example, raised bowls), then a ceramic bowl can be placed within the special bowl. Feeding bowls should be washed in hot water, using an ecological (non-toxic) liquid if necessary.

WHAT YOUR DOG NEEDS AND DOESN'T NEED*

Advised	Not advised
Wholesome, fresh, raw meat	Pig products of any kind
Occasional raw knuckle bone	Grain starch
Raw grated or liquidized vegetables	Carrots that are non-organic
Boiled or steamed vegetables	Pasteurized milk or products
Boned cooked fish	Grapes
Seaweed and kelp	Chocolate
Organic eggs	
Ceramic or pottery bowls	Stainless-steel or plastic bowls

*See: www.naturalfeeding.co.uk/dogsnf.htm

See also
Supplements (22–23), Natural Recipes (28–33)

SUPPLEMENTS

'Supplement' is a term that usually indicates a product or a foodstuff added to the basic diet, in order to satisfy a supposed need, rectify a deficiency or fulfil a certain health purpose. These usually comprise minerals, vitamins, nutraceuticals (see below) or herbs.

A dog that is fed well on a fresh, wholesome, natural and intelligently devised diet has little need of supplements. However, you can feed supplemental brewer's yeast or moderate amounts of garlic – for instance, to deter fleas. If the diet is not considered satisfactory, then omega-3 and omega-6 fatty acids, brewer's yeast, bonemeal (or minced bone), seaweed (for example, kelp), cod-liver oil and linseed oil are excellent sources of additional oils, minerals and vitamins.

NUTRACEUTICALS

Nutraceuticals – nutritional supplements with a supposed medical benefit – may be recommended if a dog is suffering from a chronic disease, such as arthritis. Products such as MSM (methylsulfonylmethane, usually derived from a by-product of the wood-pulping industry), chondroitin (normally from pig cartilage, bovine cartilage or shark cartilage), glucosamine (usually derived from the exoskeleton of crustaceans) and green-lipped mussel (*Perna canaliculus*) are often recommended for such disease situations. However, they should not be necessary if a diet has been well devised in the first place, and can be quite expensive. It has also been difficult to prove the efficacy of these products. Sometimes commercial needs replace science in the marketing of nutritional and nutraceutical products.

▲ *Moderate amounts of garlic can deter fleas.*

ANTI-CANCER AND OTHER SUPPLEMENTS

If a dog is diagnosed with cancer, there is a wide range of products with possible benefits. Antioxidant and vitamin supplements (for instance, Vitamin A, Vitamin D, Vitamin E, Vitamin C, selenium), garlic, echinacea, various other herbs, co-enzyme Q10, capsicum, various patent teas, lycopene, enzymes and other supplements are available. Readers should research the literature and Internet sources themselves for guidance on their use. It is not possible to quote unequivocal research supporting the efficacy of such products, and each animal will respond in an individual way. This book is not the forum for a debate of the issues surrounding such complex supplementation, but your vet can advise you appropriately.

In heart disease, Vitamin E, Vitamin B6, folic acid and antioxidants may often be recommended. Again, there is conflicting evidence about their value and readers should research the available information before deciding on a course of action.

If a dog is anaemic or has suffered blood loss, then a supplement rich in iron is logical. However, elemental iron can be unpalatable and may cause nausea. Foods that are naturally rich in iron should be considered, such as beetroot, black olives, liver, red cabbage and red-leafed beet.

HERBAL SUPPLEMENTS

There are herbal supplements marketed for use in a variety of chronic disease situations. These may contain Western herbs, Native North American herbs, Ayurvedic herbs or Chinese herbs. These products are usually formulated according to the general properties and virtues of the herbs they contain. They may be beneficial in a proportion of dogs. This book, however, advises that herbs of all derivations should be formulated for a specific individual patient, thus ensuring a greater likelihood of benefit. A vet skilled in the use of herbs should be consulted.

In general, supplementation can be a force for imbalance in the diet just as easily as a force for balance, and great caution should be exercised. The best initial defence is to feed a fresh, wholesome, species-suitable diet in the first place. This is possibly the best insurance against chronic disease.

▲ *Beetroot is naturally rich in iron.*

See also *Diet (20–21), Food Packaging and Labelling (24–25), Herbs (48–49), Chest and Heart Problems (76–77), Cancer: An Appraisal (88–89)*

FOOD PACKAGING AND LABELLING

Labelling laws differ from country to country; there is no world standard. In the USA, pet-food labelling is regulated at both the federal level and the state level, while in Europe each country has to implement EC directives in its own legislation.

US federal regulations, enforced by the CVM (Center for Veterinary Medicine) of the FDA (Food and Drug Administration), establish standards that are applicable for all animal feeds:

- Proper identification of the product
- Net quantity statement
- Manufacturer's address
- Proper listing of ingredients.

Some states also enforce their own labelling regulations. Many of these have adopted the model established by the AAFCO (Association of American Feed Control Officials). These regulations cover aspects of labelling such as the product name, guaranteed analysis, nutritional adequacy statement, feeding directions and declaration of the calorific content.

▲ *By law, food labels have to contain certain information.*

READ THE LABEL!

When purchasing manufactured food, it is important always to read the small print on the label. The name of the product is not enough information on which to base a purchasing decision.

DECODING THE INGREDIENTS

Ingredients are listed in descending order of inclusion rate. The term 'meat' must refer only to striated muscle, with or without the accompanying and overlying fat and the portions of skin, sinew, nerve and blood vessels that normally accompany the flesh. If the label states 'meat meal', however, this can refer to a rendered product from mammalian tissues.

If a product is named 'rabbit for dogs' or 'lamb for dogs', it must contain at least 95 per cent rabbit or lamb in the pack. If two or more major ingredients are in the name, then the combined inclusion of the named ingredients must be no less than 95 per cent.

If the ingredient name is qualified by another term – for example, 'rabbit dinner for dogs'– then it must contain at least 25 per cent of that ingredient. Similarly, for two or more ingredients, the combined inclusion should be 25 per cent or more. The ingredients list will identify other items, which should be carefully checked, especially if a dog has a known allergy to a particular ingredient.

A manufacturer can only highlight another ingredient on the label (say a herb or cheese) if it comprises at least 3 per cent of the final formulation. And if the food is labelled 'rabbit-flavour', the writing for both words must be in the same size, style and colour.

The water content of a product is important and can affect its feeding value ('meat' contains about 75 per cent moisture, 'meat meal' only about 25 per cent). A food may contain a maximum of 78 per cent moisture, unless it is labelled 'in gravy', 'in sauce' or 'stew'. The water content affects the proportions of other ingredients, so that a product with more moisture may contain a higher proportion of protein than a drier food, despite a much lower declared percentage of protein.

If labelling implies that a food can be used on its own – for example, 'complete' or 'balanced'– then it must comply with the AAFCO Dog Food Nutrient Profile. *The claims 'natural', gourmet' and 'premium' have no legally controlled status.*

Minor ingredients are often vitamins and minerals and may include artificial colours, stabilizers or preservatives. Ethoxyquin (an artificial antioxidant) is still permitted in pet food.

Regulations can only protect the consumer to a certain extent, and it is impossible to legislate for every eventuality. It must also be remembered that each dog is an individual and may not suit a particular manufacturer's product. Buying fresh food from a known source and preparing it at home offers protection against the pitfalls of the system, and fresh food should be healthier than processed food. However, variety and moderation are by-words to ensure a balanced diet over time.

See also *Diet (20–21), Supplements (22–23), Natural Recipes (28–33)*

▼ *Fresh food does not suffer from denaturing by processing.*

NATURAL WEIGHT CONTROL

In general, it is fair to state that overweight results from excessive food intake and underweight results from not eating enough. However, there are other factors that can impinge on body weight.

Obese animals are not seen in the wild. Emaciated or very thin animals are likewise not common, except in cases of disease or famine. This implies that both fat and thin animals die quickly, are killed by their fellows or do not exist. The normal wild animal is adequately nourished, with well-developed musculature.

The reasonable conclusion is that a 'wild' lifestyle and diet are conducive to healthy bodies. Conversely, domestication brings with it the possibility of unsuitable diets and incorrect exercise patterns, leading to unhealthy body condition.

▼ *Being overweight puts a strain on the heart, limbs, back and other organ systems.*

CONTROLLING WEIGHT NATURALLY

Experience has shown that dogs fed a healthy, fresh diet, based on the natural wild diet, will usually maintain their body weight and condition almost automatically. In the domestic situation, varying the quantity of food on offer can moderate the dog's weight.

Unsuitable diets can lead to obesity through an excess of certain components at the expense of others. Furthermore, cravings can be induced by certain ingredients of manufactured diets, leading to overeating. There is also the human element, in which owners feel the need to feed their dogs several meals a day and offer them titbits and scraps, thus exceeding a healthy intake each day.

It is recommended that an obese dog (in the absence of specific thyroid-gland problems that can lead to overweight) should be fed a natural, fresh diet, and advice should be sought from a holistic vet for a suitable feeding regimen. In particular, it is wise to remove carbohydrate foods from the diet altogether, until a correct weight has been

▲ *Thinness or emaciation results from underfeeding, malnutrition or disease – consult a vet about this.*

REMEMBER...

It is clearly undesirable to have an underweight dog and, in recent years, obesity in dogs has also been very much in the news. In either case, it is recommended that the advice of a vet should be sought. Good nutrition and exercise programmes are also vital to health.

reached. Dogs with thyroid problems, possibly induced by immune disturbance, should be taken to a vet for advice and care.

Excessively thin dogs should be checked by a vet for worms, thyroid excess, pancreatic insufficiency (again, often an immune-related problem), malabsorption or other metabolic or endocrine problems. In the absence of treatable disease, they should be fed a natural and fresh diet, with advice from a holistic vet on both regimen and quantity. If the individual dog can tolerate animal fat in quantity, pieces of hard lamb or beef fat can be fed, to assist weight gain.

EXERCISE

In addition, the exercise regimen enjoyed by the dog can affect his body weight and should be adjusted according to need. It is important to remember that, should the dog sustain an injury – thereby restricting his exercise capability – or if the dog walker cannot perform the usual walking routine, food must be reduced accordingly to prevent overweight becoming an issue.

Some conditions can lead to an impression of excess weight in a dog when there may be a medical reason that requires intervention. Conditions such as heart disease (leading to fluid accumulation in the abdomen), cancer (with a swelling in the abdomen), pregnancy and Cushing's Syndrome can all make a dog look overweight. Veterinary help should be sought to ensure the correct diagnosis.

See also *Diet (20–21), Worms (40–41), Gastrointestinal Problems (66–67), Autoimmune and Endocrine Disorders (86–87)*

NATURAL RECIPES

The argument for providing a diet as close as possible to that of a wolf or wild dog is strong (see pages 20–21). Nevertheless, it may understandably be somewhat too 'extreme' for some households. For this reason suggestions are given for feeding a wholesome and reasonably natural diet, through the recipes on the following pages. Many more ideas can be added, to create a truly varied fresh diet for your dog.

The recipes on pages 29–33 are not set out in exactly the same way as recipes for human delicacies, because a dog neither seeks nor appreciates fancy food preparation.

These recipes are not intended to make dog food look like human food, although fresh food can be sourced from the same places as the family's diet, and the preparation methods may overlap. It is recognized that cooking is not part of a truly natural diet for a dog, but is acceptable.

Organic food is recommended, in order to minimize the potential introduction of chemicals into the dog's diet. If the dog is ill, organic food becomes even more important in support of the dog's healing efforts.

CAUTION

These recipes are not designed for exclusive feeding but as part of a natural feeding programme, which should include a variety of wholesome ingredients, each in moderation.

ACHIEVING BALANCE IN YOUR DOG'S DIET

Dietary balance is not achieved in each meal, whether for humans or our dogs; it comes from offering a variety of wholesome and species-suitable foods over a period of weeks and months. The body is then able to achieve dietary balance for itself. Offering your dog a wide range of fresh, wholesome and species-suitable foods is the best route to a balanced diet. Offering just one type of food is likely to set up imbalances for which the body cannot compensate.

Supplements can be selected in order to compensate for perceived weaknesses in the range of foods offered.

GETTING STARTED

Portions can be judged according to your experience with a particular dog. A medium-sized dog will usually easily manage ½–1 kg (1–2 lb) of fresh meat in a day, along with vegetables and oils. Your dog may not object if you wish to share some of these recipes.

See also *Diet (20–21), Supplements (22–23)*

MEAT, HERB AND BONE SAUSAGE

This is suitable for a dog that may not be able to manage chunks of raw meat or whole bones.

- A portion of fresh lamb or beef, minced
- Half of that weight of fresh bone, finely ground or minced (no splinters)
- A good sprinkling of kelp
- A good sprinkling of dried herbs (*Herbes de Provence* or mixed herbs)

Roll this mixture into a sausage shape and put it down for the dog along with some boiled fresh vegetables, moistened with olive oil.

This dish is not cooked and should not be kept for long.

SUITABLE VEGETABLES AND HERBS

- Suitable vegetables for dogs include: celery, sweet potato, spinach, broccoli, carrots (must be organic), parsnips, watercress, various seaweeds (sea vegetables) and nettles, but others are also satisfactory.
- Good culinary herbs include: parsley, oregano, thyme, sage, fenugreek, garlic, turmeric and coriander.

WHOLESOME STEW

This recipe can be combined with dumplings: see page 32.

- Meat chunks (beef or lamb, without trimming off the fat or gristle)
- Carrots, celery or broccoli, finely chopped
- Turnip, swede or parsnip, chopped
- Shallots (whole)
- Kelp
- Parsley, minced or finely chopped

Mix the ingredients together and cook on the hob or in the oven until the vegetables are soft.

When the dish has cooled, give it to your dog.

▲ *Broccoli is a reliable and nutritious vegetable for dogs and very easy to prepare.*

RABBIT STEW

This recipe can be combined with dumplings (see page 32).

Note: raw wild rabbit (cottontail) may harbour tapeworm.

- Sweet potato, carrot, celery, peas, leeks, turnip, onions
- Rabbit meat (wild)
- Olive oil for frying
- Herbs (parsley, rosemary, marjoram and thyme)
- Vegetable stock (unsalted – that is, probably home-prepared)

Chop the vegetables (unpeeled), then chop the rabbit meat, without the bones (cooked bones are not recommended).

Fry the rabbit pieces in a little olive oil for about ten minutes, then put into a casserole. Sprinkle with the herbs. Add the stock (or water) and place in the oven until the meat is cooked through.

Add the chopped vegetables and return to the oven for a further 45 minutes.

When the dish has cooled, give it to your dog.

QUINOA BAKE WITH VEGETABLES

Meat can be added if desired, but should be cooked more than the other ingredients; part-cooked meat can be dangerous, even for dogs.

Note: while this recipe contains starch, the presentation of starch is different in quinoa from that in cereal grains.

- Quinoa
- Courgettes, broccoli, spinach and garlic, sliced
- Squash, chopped, but not peeled
- Olive oil for frying
- Fresh parsley, minced or finely chopped

Rinse a portion of quinoa and lightly boil it with double the quantity of water.

In the bottom of a casserole pan or dish, lightly sauté the vegetables in the oil.

Add the par-boiled quinoa and the parsley. Put on the lid and bake in the oven until the fluid has been absorbed.

When the dish has cooled, give it to your dog.

◀ *Celery and crushed celery seeds offer good nutritive qualities.*

BRAISED CHICKEN

This is a real 'protein fix'. You can also give it to your dog with boiled eggs. Pheasant could be used instead of chicken, when available (fresh road-kill is fine; beware of lead pellets if it is shot).

- Chicken pieces
- Coconut oil
- Peanut butter (unsalted) or nut butter (unsalted)
- Parsley, minced
- Garlic, chopped
- Turmeric powder

Fry the chicken pieces without any bone (cooked bones are not recommended) in a little coconut oil. When it is tender, add the nut butter, parsley, garlic and turmeric.

Cover and cook to the required level.

When the dish has cooled, give it to your dog.

▶ *Lamb, cooked or raw, makes a satisfying and nutritious element of a natural diet.*

MEAT, BEAN AND CHEESE BAKE

This meal is another great 'protein fix'.

- Chunks of beef, lamb, turkey, chicken, rabbit or pheasant
- Broad beans
- Soft goat's cheese
- *Herbes de Provence*
- Eggs, beaten

Braise or fry the meat, and boil the broad beans, then place in an ovenproof dish and mix together.

Crumble the goat's cheese over the mix and sprinkle it with the herbs.

Beat as many eggs as you need to cover the portion you have prepared, then pour the beaten eggs over the mix.

Place in the oven and bake until set.

When the dish has cooled, cut off the required portion and offer it to your dog. Stand back!

Keep the remainder in the refrigerator for another day.

VEGETABLE PENNE COMBO

This is a moderately high-starch recipe, not for the wheat-intolerant dog. The provision of pasta or other starch is only recommended for 'high-energy' dogs who can burn off the 'empty calories'. It is most definitely not for a dog that needs to lose weight. If a dog has cancer, starch is best avoided altogether.

- Vegetables and herbs of the season (both root and leaf), such as celery, parsley, coriander, basil, watercress and garlic, diced
- Potato peelings (optional)
- Vegetable or potato water (unsalted)
- Wholemeal penne pasta
- Olive oil (or sunflower-oil margarine)

Place the diced vegetables and potato peelings (if available) in a large saucepan and cover with any previously saved vegetable or potato water. Boil until just starting to become tender.

Add the pasta and continue to boil until it is soft (that is, beyond *al dente*), ensuring there is sufficient water to prevent the pan running dry. Drain the liquid and stir in the oil or margarine to prevent clogging.

When the dish has cooled, give it to your dog. This will last several days in the refrigerator. It can be served as a meal in itself; because of the starch element, it is best fed separately from a meat meal.

TASTY AND NUTRITIOUS DUMPLINGS

The raw dumplings can also be added to a stew or casserole for the last ten minutes of its cooking time.

- Gram flour
- Shredded beef or lamb suet (using vegetarian suet will make a vegetarian version of this dish)
- Herbs of your choice (parsley is excellent), dried or fresh minced, according to season
- Turmeric powder (if your dog is a real curry lover, a little cumin, coriander and cardamom powder can be added, for extra flavour)

Mix together the gram flour and suet (experiment with the proportions, to get the texture you desire), then add the herbs and turmeric.

Mix in just sufficient water to make a workable dough. Form into separate balls of whatever size you wish (smaller ones will cook more quickly).

Boil for about ten minutes.

When the dish has cooled, give it to your dog.

ORGANIC LIVER CAKE

For use as a training aid or as treats. Adapted from a recipe kindly submitted by a client of the author.

- 500 g (1 lb) organic liver (lamb or calf liver)
- 1 whole organic garlic bulb
- 1 teaspoon fresh or dried herbs (parsley, oregano, sage, thyme), finely chopped
- 170 g (6 oz) organic jumbo or porridge oats

Put all the ingredients into a food processor and blend until thoroughly mixed. The mixture should be thick and heavy and not too wet. If it is too wet, add a few more oats.

Spread into a greased baking tin, to about 6 mm (¼ inch) thickness and cook for about 15 minutes. It will be just crusty on the top, but pliable enough to divide easily into pieces.

Once cooled, cut or break into pieces about 5 cm (2 inches) square. These can be frozen and one can be defrosted each day to give as a treat, or to take in your pocket on your walk or training session.

CHEESY MORSELS

This can also be made with crushed cranberries or blueberries in place of the herbs, for a special health-giving treat.

- Cottage cheese or soft goat's cheese
- Gram flour
- Sesame flour
- Herbs (such as chives, rosemary, garlic and parsley), finely chopped or grated

Mix all the ingredients together thoroughly and break into bite-sized pieces.

Cook lightly in the oven until they retain their form.

When the morsels have cooled, give some to your dog.

▲ *Minced fresh parsley is nutritious and can help maintain the health of the urinary system.*

VACCINATION AND ALTERNATIVES

The normal 'core' puppy vaccinations (compulsory by law in many places) consist of components against distemper, hepatitis (adenovirus) and parvovirus. Rabies is also a core vaccination, but is described separately. Non-core vaccinations on offer are for lyme disease, leptospirosis (two forms) and kennel cough.

There are several serious diseases that can kill your dog, making some form of protection essential. The main ones are:

- Distemper
- Hepatitis (adenovirus)
- Parvovirus
- Leptospirosis (two forms): *L. canicola* and *L. icterohaemorrhagiae*
- Rabies.

There are other diseases of dogs (usually non-fatal) for which vaccines are available. A nosode (see opposite) is available for each of the above diseases, so ask your holistic vet.

DISTEMPER

This is a commonly fatal disease affecting all body systems, resulting in great suffering. When the nervous system is affected, dogs

can exhibit convulsions. It is spread by contact and infected discharges.

ADENOVIRUS (INFECTIOUS CANINE HEPATITIS)

This disease can be fatal, but even those dogs that recover can be very ill and may never regain full health. It is spread by infected saliva, faeces and urine.

PARVOVIRUS

This is a commonly fatal viral disease affecting the gastrointestinal tract. It presents as violent vomiting and diarrhoea with dysentery. Dehydration, heart damage, collapse and death usually follow, despite intensive care.

LYME DISEASE

This is a tick-borne bacterial disease associated with various *Borrelia* species of bacteria. It affects the musculoskeletal system, causing lameness and also sometimes heart damage.

LEPTOSPIROSIS

There are two forms of leptospirosis that affect dogs: *L. canicola*, which damages the kidneys, and *L. icterohaemorrhagiae*, which affects the liver, with jaundice. In the case of the latter, infected urine from rodents (even just in a body of water) is the usual source of infection for dogs. In the case of *L. canicola*, other dogs and wild canines are the reservoir of infection.

◀ *Your dog may contract parvovirus from sniffing infected material while exercising.*

NOSODES

Nosodes are medicines prepared according to the same methodology as normal homeopathic medicines, but they are derived from disease material – whether tissues, discharges or secretions. Because of the extreme dilution process (usually to the thirtieth centesimal dilution or 'potency'), no active material remains from the original infection, so they are safe. Nosodes serve many different purposes in homeopathy, one of which is their prophylactic use in an attempt to prevent infectious disease. Nosodes have been made from most infectious canine diseases, and new ones can be made to order by a pharmacy with the correct experience and equipment.

Many dog owners in the USA and other countries rely on protection by nosodes alone for the 'core' diseases mentioned above (with the exception of rabies, for which conventional vaccination is compulsory in most states of the USA). See page 36 for discussion on nosodes.

KENNEL COUGH (CANINE TRACHEOBRONCHITIS)

This disease is a rarely serious, highly infectious respiratory disease, presenting mostly 'nuisance' symptoms that can often last for three weeks. It has a complex aetiology, and vaccines are variably effective. Intra-nasal vaccination is not well accepted by dogs.

THE VACCINATION DEBATE

There is a great deal of discussion about the necessity, advisability and safety of vaccination and about the efficacy of the alternatives (nosodes). The Internet will provide the reader with much to ponder when making a decision. Certainly there is no scientifically accepted proof of efficacy for nosodes as a means of protection against infectious diseases. Equally certainly, no manufacturer can claim 100 per cent safety for vaccine products, and some people believe there are serious dangers in using them on dogs.

Vaccine products contain many more components than the required antigenic material. The list of possible 'other' ingredients includes mercury, aluminium, phenol, formaldehyde, antibiotics, oils, animal tissues and even the possibility of cancerous DNA from continuous cell lines on which the viruses are cultured.

Science provides no definitive answer to how often vaccine booster shots should be given. There is no science supporting an annual booster and the debate goes on: whether to revaccinate every second year, every third year, intermittently or once only.

Antibody titre-testing (to ascertain the level of antibodies in the blood) gives only a very incomplete view of immunity. The presence of circulating antibodies indicates a level of immunity, but a negative result does not necessarily indicate a *lack* of immunity. This test is often recommended to help decide whether revaccination is necessary, but it cannot serve as a complete guide.

There are many anecdotal reports of the side-effects of vaccination, with conditions

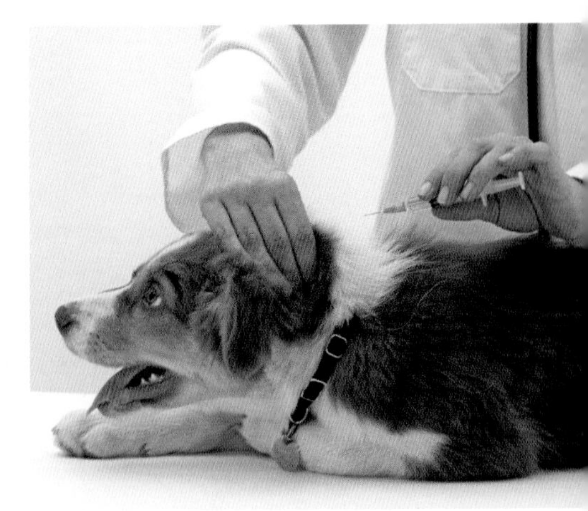

▲ *Vaccination is the conventional way to protect your puppy or dog from disease.*

such as eczema, epilepsy, autoimmune disorders, colitis, tumours and cancer being cited (some including fatality). These far outnumber the officially reported cases. In a survey of more than two hundred cases conducted by the author, in those animals in which the start date of a chronic disease was certain, it occurred within three months of a vaccine event in 80 per cent of cases.

Many dogs are given nosodes as the sole means of protection, and have withstood definite local outbreaks and contact with the named diseases. There are reports of high titres against parvovirus in unvaccinated dogs that have been given nosodes only, never having received a conventional vaccination against the disease. This can only arise as a result of infection that has been encountered and successfully withstood.

The protection of precious pets against dangerous infectious disease is vital. It is the duty of each reader to research the available information, whether in books or on the

Internet, to help when making a decision on how to protect your dog.

RABIES

Rabies is a fatal viral infection of dogs, and humans can also be infected. It is therefore a 'zoonosis' – a disease of animals that is transmissible to humans, among whom it is potentially fatal.

The usual mode of spread is by a bite from an infected animal. In 2007 the US Center for Disease Control (CDC) formally declared that canine rabies had ceased to exist in the USA. 'The elimination of canine rabies in the United States represents one of the major public-health success stories in the last fifty years,' stated Dr Charles Rupprecht, chief of the CDC rabies programme. However, it is still possible for dogs to be infected by a bite from an affected wild animal, such as a bat, coyote, skunk, raccoon, wolf or fox. A domestic dog is considered most likely to infect humans. For this reason, most states in the USA have laws that make rabies vaccination of dogs compulsory; these laws may vary locally, but cannot be less stringent than the respective state legislation. The regimen that is specified by law can also vary, from once every year to once every three years. The legal starting age for vaccination may also vary, usually from eight weeks to six months.

In Europe, rabies is still present, although large areas of western Europe are gaining rabies-free status, helped by the strict implementation of regulations controlling the movement of pet animals and the necessity of vaccination before travel. The main sources of rabies infection are dogs in eastern European countries and on the borders of the Middle East, foxes, raccoons and insectivorous bats.

There is no alternative to legally required vaccination shots. However, it may be that a puppy or dog can be given a 'nosode' and other homeopathic medicines by a holistic vet, in an attempt to minimize any possible ill effects from the rabies shots. No advisor can make the decision for you.

▼ *A wild animal, such as a raccoon, can infect your dog with rabies, via a bite.*

See also *Homeopathy (52–53)*

PARASITES: FLEAS

The dog flea, *Ctenocephalides canis*, is an insect parasite, with one pair of legs being very powerful, allowing it to leap about 15 cm (6 inches). It feeds on the dog's blood, but it breeds off the dog, in warm, sheltered, dark nooks and crannies.

▲ *The dog flea* (Ctenocephalides canis) *breeds in warm, dark places, off the dog.*

The flea larvae are about 4 mm (1⁄6 inch) long. They feed on accumulations of dry blood and organic substances in corners and crevices. When infestations are very heavy, the combination of greyish larvae and white eggs give the 'nest' area a characteristic 'salt-and-pepper' appearance. Flea breeding is more active during the summer months, but modern heated houses can encourage year-round breeding. Other species of flea can also attack dogs and foxes, and domestic cats can be a reservoir.

The adult flea is about 2.5 mm (1⁄10 inch) in length and is narrow relative to its height. A dog may scratch and nibble when it has a flea or fleas in his coat, on account of the irritant nature of a flea bite. Fleas may congregate on the head, neck, shoulders, sacrum and tail head, but can feed anywhere on the dog. They can be very well hidden on dogs with a long coat. Sometimes a flea can be seen scurrying across the hairless part of a dog's abdomen.

Only in very severe infestations can a dog's vitality be threatened, but fleas can capitalize on a debilitated dog and may, even when as few as one or two are present, set up allergic reactions and skin disease, as a result of the nature of flea saliva. While a few fleas on a dog hardly represent a health threat, either to humans or animals, neglect of a flea problem can lead to serious house infestation, as a consequence of the rapid reproductive cycle of fleas.

COMBATING FLEAS

Flea-control programmes must be carried out both on the dog and in the house.

On the dog, aromatherapy oils can offer a good first line of defence and an effective deterrent, if the threat is not massive. Cedarwood, eucalyptus, garlic, lemongrass, lemon, neem and pennyroyal are well-known insect and flea repellents, although pennyroyal is not recommended on account of its abortifacient (abortion-inducing) properties. These oils should be diluted in water (two or three drops per half-pint of water) and then combed through the dog's coat.

Feeding a dog garlic, lemongrass or brewer's yeast will also act as a deterrent, as fleas do not like those flavours. Combing a dog with a flea comb or electric flea comb can also be an effective method of control.

In the house, chemical options are available. Natural options are aromatherapy oils (see above), strategically placed in areas where a flea might choose to breed (such as carpet edges, skirting boards, floorboard cracks and down the sides of chairs). Crushed chrysanthemum, fleabane or *Tagetes* (marigold) may also act as a deterrent. There are also some proprietary products of natural origin that are sold for flea control both on the dog and in the home.

REMEMBER...

- Fleas are a constant threat in dog households. Modern wisdom tends to recommend powerful chemicals to control this threat, but there are more natural and ecologically friendly methods available.
- Fleas can bite humans, if hungry.

If a severe flea infestation occurs, resorting to strong manufactured chemicals (sourced from your vet) may be necessary, with reversion to more natural methods once the situation has been brought under control.

▼ *Modern conventional flea-killing compounds are often delivered by the spot-on method.*

See also *Vaccination and Alternatives (34–37), Worms (40–41), Aromatherapy (54–55), Skin Problems (68–69)*

PARASITES: WORMS

There are six different types of worm that can infect your dog: roundworm, tapeworm, hookworm, whipworm, heartworm and lungworm.

ROUNDWORM (*TOXOCARA CANIS, T. LEONINE*)

This parasite inhabits the intestines, usually in puppies, and produces a pot-bellied appearance in the pup, with a dry, scurfy coat. Loss of weight and condition and eventual emaciation are more advanced signs.

Puppies are infected *in utero*, as a result of larvae migrating through a pregnant mother's tissues into the womb. All puppies should be assumed to be infected from birth. Your vet will advise on an appropriate deworming programme.

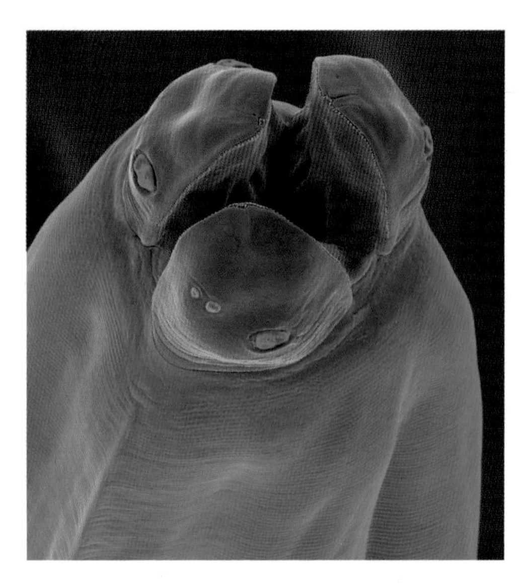

▲ *Roundworms (*Toxocara canis*) commonly infect puppies* in utero *(before birth).*

Unless a puppy or dog vomits an adult worm or passes one in faeces, clinical signs or microscopic examination of the stool for eggs are the only methods of detection.

Larvae have been known to infect humans, and it is therefore important to keep your dog clear of roundworm. Regular laboratory examination of faeces is sensible, especially if there are babies or toddlers in the household.

TAPEWORM (*DIPYLIDIUM CANINUM*)

The tapeworm is a ribbon-like parasite of the intestines. It is composed of segments that carry eggs, and these break off and are passed via the anus. They can be seen with the naked eye, like moving grains of rice around the anus or in the stool. The parasite can be a serious competitor for food, and an affected dog can lose weight very rapidly.

Intermediate stages in the tapeworm lifecycle are harboured by small prey animals, such as mice and cottontail rabbits. Fleas can also carry the infection, which means that a dog with fleas may also have tapeworms.

HOOKWORM (*ANCYLOSTOMA* AND *UNCINARIA* SPP.)

A common parasite of the dog's intestines, these very slender worms are about 12 mm (½ inch) long. Despite their small size, a large number of them can take a significant

▲ *Tapeworms (*Dipylidium caninum*) are picked up by eating cottontail rabbits and other small creatures.*

proportion of the dog's blood per day.

Eggs from a dog's faeces settle on the ground, and the larvae hatch and survive on moist vegetation or soil. They can infect by means of ingestion or via the skin. In addition, like roundworms, hookworms may infect a puppy *in utero.* These worms may also infect humans.

WHIPWORM (*TRICHURIS VULPIS*)
Whipworms are microscopic worms inhabiting the colon of dogs. A heavy infestation can cause anaemia. The stool may also contain blood or mucus. A dog can pick up these worms by taking in infected food or water.

HEARTWORM (*DIROFILARIA IMMITIS*)
Heartworms are small, parasitic, threadlike roundworms, which travel through the bloodstream and, in the final stages of life, reside in the heart of their host. They can destroy heart muscle, resulting eventually in heart failure, and can cause serious heart disease in dogs if an infection goes untreated. Infection occurs via the bite of a mosquito.

LUNGWORM (*ANGIOSTRONGYLUS VASORUM*)
This parasite affects the lungs and is acquired by eating slugs or snails.

TREATMENT
It is sensible to consult your vet about the treatment and prevention of these various worm infestations. Conventional chemical dewormers are generally very effective, if correctly chosen. There are natural dewormers on the market, but there is no published proof of their efficacy. Herbs that are well known for their reputed anthelmintic (worm-destroying) action are wormwood, wormseed, lad's love, southernwood, santonin and cat thyme (for roundworms), and pomegranate and male fern (for tapeworms). It is recommended to seek the opinion of your holistic vet, if you wish to try natural dewormers. In the case of those parasites that have migratory stages within the dog's tissues, there is a chance that a nosode (see page 35) could be made and may help to prevent or treat infestation. This is a poorly researched area, but your holistic vet should be willing to discuss the options with you.

See also *Vaccination and Alternatives (34–37)*

PARASITES: LICE, MITES AND TICKS

Lice, mites and ticks are all ectoparasites of the dog's skin – they live on the outside of the host. Infection is usually by close contact with infected animals, except in the case of ticks, which jump onto an animal as it passes by the herbage on which it lives.

LICE (*TRICHODECTES CANIS* AND *LINOGNATHUS SETOSUS*)

Lice are biting or blood-sucking insect parasites that are visible to the naked eye. However, they are very tiny and flattened dorso-ventrally, so seeing them can present a challenge. If your dog is itchy, losing hair and has a poor coat, a vet or dog groomer can help you to check for lice.

Lice spend their entire lives on the dog, so infection is by direct contact with an infected dog. Lice lay eggs (termed nits) on the hair shafts. The lifecycle takes about 21 days to complete, and the eggs may be visible attached to a shaft of hair.

▲ *Lice are blood-sucking or biting insect parasites.*

MITES (*SARCOPTES, DEMODEX* AND *OTODECTES CYANOTIS*)

Three types of mite can affect your dog: sarcoptic (or fox) mange mite, demodectic mange mite and the ear mite.

Demodectic mange

This mange is not usually itchy, but areas of skin – especially around the muzzle, eyes, ears and feet – take on a grey and thickened appearance. Skin infection may follow, with resultant pustules forming. Hair is usually absent from the worst-affected areas. Infection is generally via close contact with an infected dog, but many dogs carry this disease without showing any signs. Anything that can disturb the immune balance may trigger the disease in a carrier dog.

Sarcoptic mange

This is an extremely itchy condition, with tufts of hair being ripped out through scratching and rubbing. The dog's coat looks very unkempt and sparse. Infection usually occurs from close contact with an infected dog or fox.

Ear mites

Ear mites feed on the lining of the ear canal and cause great irritation, leading to head-shaking and scratching or rubbing the ears. A characteristic brown wax discharge can usually be seen. Dogs are usually infected by close contact with an infected dog or cat.

TICKS (*IXODES* SPP.)

Ticks are blood-sucking parasites that spend most of the year in rough herbage, jumping onto a warm-blooded animal as it passes by the herbage. Some ticks feed in the spring and others in the autumn. They are tiny when first attached and grow as they fill with blood. They attach themselves by burying their mouthparts into the dog's skin, making removal a skilled job. If a tick is incorrectly removed, a stubborn lesion can develop at the site of attachment. Ticks are significant in that they can spread viral diseases, most notably lyme disease (see page 35).

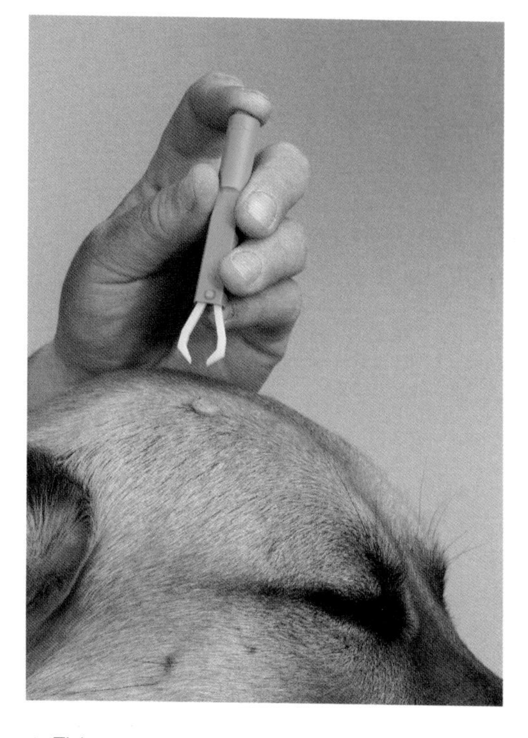

▲ *Ticks are acarids that feed on the dog's blood.*

TREATMENT AND CONTROL

Regular grooming provides a very useful opportunity for close inspection of your dog's skin and coat. Combing and brushing remove any debris and can also remove parasites or parasite eggs.

Lice are usually easy to eliminate and may respond favourably to regular application of diluted tea-tree oil or neem. There are also chemical agents that your vet can supply.

Ear mites may similarly be controlled quite simply, using diluted aromatherapy oils such as tea tree and neem, a powder or chemical agents supplied by your vet.

Sarcoptic mange and demodectic mange can prove very difficult to clear. Sometimes the best treatment is a combination of aromatherapy oils, homeopathic Sulphur and chemical treatments.

Ticks are usually carefully removed using a special tool.

CAUTION

In all cases, it is recommended that you should seek the advice of a vet or holistic vet before embarking on treatments.

See also *Grooming (18–19), Vaccination and Alternatives (34–37), Worms (40–41)*

NEUTERING: SPAYING

It is commonly recommended that bitches should be neutered (spayed), if they are not required for breeding. There are, of course, arguments both for and against spaying, which is also known as ovarohysterectomy.

Most bitch puppies reach puberty at about six months of age (5–18 months). They commonly show a five- to eight-monthly cycle from then on, which may wane with age. Bleeding at the start of a 'season', stopping just before ovulation, occurs from the lining of the vagina; it is not analogous with human female bleeding. Regardless of mating, the 'season' will cease soon after this and the bitch then goes into pregnancy if successfully mated, or 'pseudo-pregnancy' if not. The latter lasts for a full nine weeks, as in a normal pregnancy, whether it is noticeable or not. There is then a 'nesting and nursing' phase, which again may or may not show.

If a bitch is not spayed, there can be heavy blood spots during the early season. She may become wayward during seasons, seeking a mate. She may be moody, clingy or morose during her pseudo-pregnant phase. She may become obsessed with nest-making or with a 'surrogate puppy' during the false nursing phase, and may produce milk. She may, of course, fall pregnant, if she meets a dog at the right time. Male dogs in the vicinity, during the bitch's season, may damage property or give rise to dangers on the road in attempts to get to her, or may even suffer health problems themselves.This can be a disaster if the dog is of too large a type, or if unwanted puppies result.

Some bitches will produce mammary tumours (usually benign), which grow cyclically at each 'season', and others may develop pyometra (inflammation of the

◀ *A bitch can become obsessed with a cuddly toy during hormonal phases.*

THE GENITO-URINARY SYSTEM: BEFORE AND AFTER SPAYING

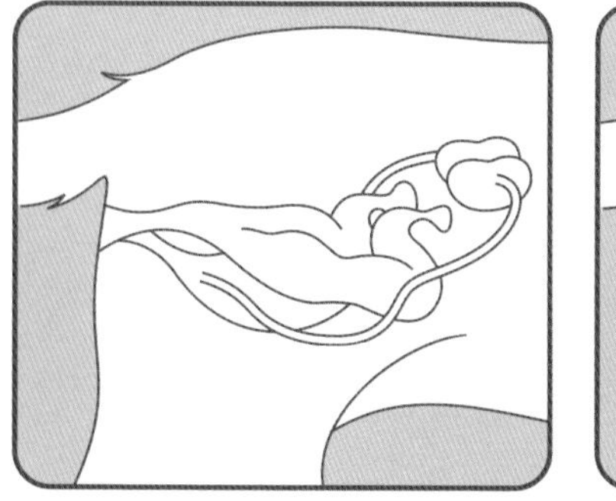

Normal

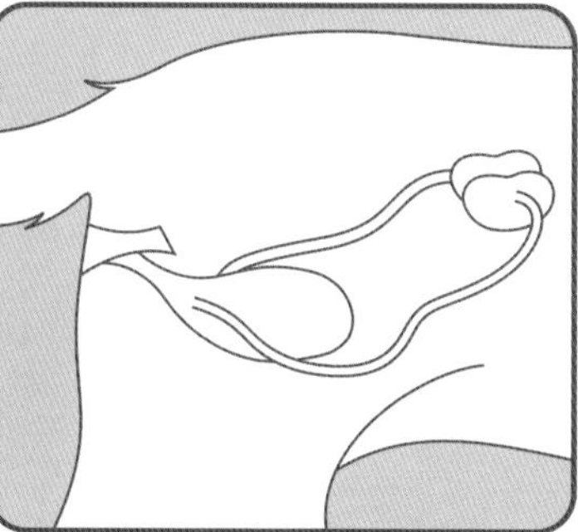

Spayed

uterus). These diseases are rare, especially when the bitch has natural holistic feeding and management.

WEIGHING UP THE OPTIONS

Disadvantages (apart from surgical or anaesthetic risks) can result from hormone imbalances and structural changes, as a consequence of removal of the womb and ovaries. Weight control can also be a problem – the bitch may become greedier. And the coat, especially in Spaniels and Setters, may become pale, wispy and profuse; the downy undercoat can be lost and bare patches may appear on the flanks. Very rarely, disastrous 'false pregnancies' occur, with massive milk production, each six months.

A common 'side-effect' is urinary incontinence (enuresis), which may be slight or extreme. It can vary in intensity, following the previous six-monthly pattern, and usually occurs during sleep or resting.

If a bitch cannot be controlled, thus risking unwanted pups, the only options are spaying or heat-suppression by means of injection. The latter can carry an increased risk of mammary cancer, the injection may be painful and the hair may change colour at the injection site. Birth-control injections ('morning afters') are potentially bad for the bitch and cannot be repeated in any one season; they may also prolong the season. The 'mutilation' that spaying represents is understandably undesirable to many, but may be a 'necessary evil' to counteract the unnatural situation of the domestic bitch, away from her ancestral 'pack' context.

The operation must be performed about 12 weeks after a season, away from her times of greatest hormonal activity. Some advocate spaying before puberty, and the debate on this continues. Hormone supplements after spaying are generally not considered safe.

If a bitch does not suffer with seasons and can be well controlled, then spaying is not essential. Chlorophyll tablets, which effectively mask or disguise a bitch's scent, can prevent the antisocial aspects.

See also *Natural Weight Control (26–27), Reproductive-system Problems (80–81)*

NEUTERING: CASTRATION

It is commonly recommended that male dogs should be castrated, if they are not required for breeding. This is a thorny issue for many dog owners and there are, of course, arguments for and against this practice. When trying to arrive at a decision, it helps to be objective, and the issue is therefore discussed here.

Most young male dogs reach puberty at about six months of age. They may or may not show overtly sexual behaviour at that time. In some dogs, their behaviour can become a real nuisance and embarrassment. It is not clear what factors lead to normal sexual expression becoming antisocial, unacceptable or even a health risk to the dog himself. In some cases, it can become all of these things.

One possible factor is the absence of the natural ancestral pack environment. This is common to all domestic dogs, however. It is to be hoped that a good, solid, loving home, with security and a naturally 'dominant' human pack leader, will usually lead to the expression of male-type behaviour remaining at very acceptable levels.

Another factor may be diet. Again, it is reasonable to expect that a natural, fresh, wholesome and varied diet, without chemical additives, will encourage normal health and behaviour.

PROS AND CONS

Because castration is, after all, a form of mutilation, it is arguable that it is better not to subject your dog to this procedure if you don't have to. It is then important to ensure that your 'entire' male does not create unwanted puppies – there are too many of those in the world already.

An entire male can also suffer, both mentally and physically, if he is walked where 'on heat' bitches have recently been. The residual scent will be easily detected by his sensitive nose. Walks can then become a nightmare.

Hypersexual dogs can also be a tremendous nuisance to neighbours, severely damage property and be a danger on the roads. Care is needed on all these fronts.

REMEMBER...

Castration is rarely indicated for health reasons. However, there are also social and domestic considerations that may affect the owner's decision – but no surgery or anaesthetic is without risk.

While it is often argued that no possible disadvantages ensue from this surgical intervention, there are some potential side-effects.

THE GENITO-URINARY SYSTEM: BEFORE AND AFTER CASTRATION

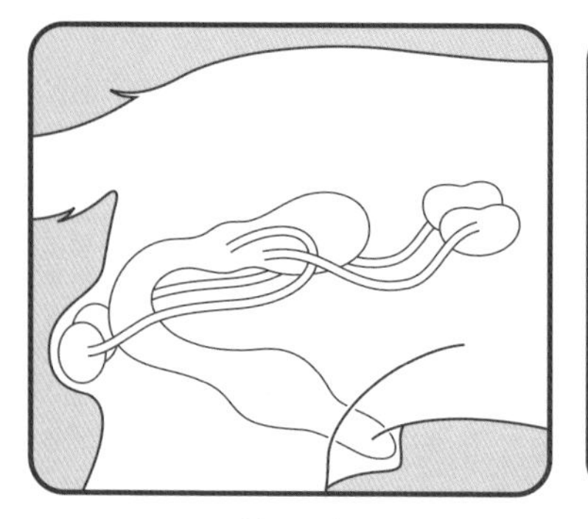

Normal

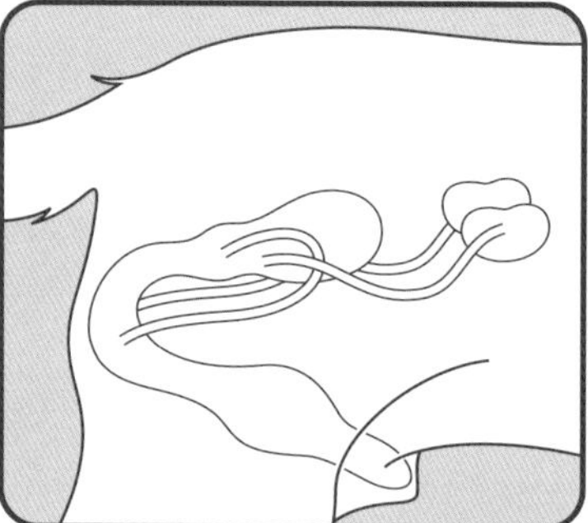

Castrated

Should problems arise, in many cases, herbal and homeopathic treatments can help regain your dog's balance, without recourse to surgery. There is also a drug treatment that may help to moderate male hypersexuality.

Should normal sexual behaviour transcend the acceptable and persist for a considerable period, despite treatment, castration may be seen as the lesser of two evils. If it is not carried out, life may become stressful for both dog and family, and health problems such as prostate enlargement become more likely.

In cases of cryptorchidism (retained testicle), castration becomes a necessity, preferably before the age of three. If it is not done, experience suggests that cancer is an extremely likely outcome.

It is often argued that castration brings no ill effects. However, in some dogs the coat can change, fat distribution in the body may alter, weight control may become more difficult and even a dog's character can change.

Whether castrated or not, a dog will be healthier on a fresh and varied diet and will benefit from essential fatty acids, such as in evening primrose, star-flower or fish oils. Occasional cautious zinc supplementation can also be valuable for the dog's ongoing health. If he suffers any of the side-effects of castration, as mentioned above, homeopathy can sometimes help.

▲ *Walking an entire male can be difficult if he detects an 'on-heat' bitch in the vicinity.*

See also *Natural Weight Control (26–27), Homeopathy (52–53), Reproductive-system Problems (80–81)*

HERBS

Herbal medicine is the oldest medical practice known to humans. Ancient texts from China, Egypt, Greece, the Arab world and medieval Europe include information on the medicinal use of herbs, which is the direct ancestor of modern Western medicine.

▲ *Dandelion (*Taraxacum officinale*) is a potent natural diuretic and rich in Potassium.*

Many modern drugs have been derived from plants, and all human cultures appear to have an active herbal tradition, with a deep understanding of the value of herbs in maintaining health and treating injury and disease. There is evidence that animals also use plants for medical reasons (zoopharmacognosy), and therefore herbal lore may be partly instinctive, even in humans.

In herbal medicine, it is not always possible to distinguish between the nutritional benefits of a plant and its medicinal properties. In medical practice, whether for humans or animals, there should not, in any case, be a separation of medicine and nutrition.

PLANT GROUPINGS

- Plants contain different classes of medically active substances, in unique combinations, including alkaloids, glycosides, saponins and flavones. This can provide one basis for classification.
- Herbs can also be grouped according to their general action, for example: alterative, anodyne, anthelmintic, antibacterial, anticatarrhal, anti-emetic, antifungal, anti-inflammatory, antilithic, antispasmodic, aperient/laxative, aromatic, astringent, bitter, cardiac, carminative, cathartic/purgative, cholagogue and anti-cholagogue, demulcent, diaphoretic, diuretic, ecbolic, emetic, emollient, expectorant, febrifuge, galactagogue, hepatic, hypnotic, nervine, rubefacient, sedative, sialogogue, soporific, stimulant, styptic, tonic, vesicant and vulnerary.

HOW HERBS ARE USED

The skilled herbalist, whether practising Western herbal medicine, Traditional Chinese Medicine (TCM), Ayurvedic medicine or some other tradition, will choose individual herbs or a combination to obtain the precise effects required for the patient.

Herbs can be given as capsules, tablets, dry powder, 'tea', alcoholic tincture or

COMMON HERB ACTIONS

Herb	Action
Angelica	Febrifuge (helps reduce fever)
Black cohosh	Antispasmodic (relaxes muscle)
Burdock	Alterative (produces a general healing tendency)
Comfrey	Demulcent (coats and soothes mucous membranes)
Dandelion	Diuretic (promotes urine flow)
Elder	Diaphoretic (promotes sweat)
Elecampane	Tonic (generally 'strengthens' the body)
Flax seed	Aperient (promotes defecation)
Garlic	Anthelmintic (discourages internal worms)
Golden rod	Astringent (has a drying action)
Hawthorn	Cardiac (acts on the heart)
Hops	Nervine (has a healing benefit on the nervous system)
Horseradish	Stimulant (boosts metabolism and excites nerve pathways)
Marigold	Vulnerary (helps wound and injury healing)
Sage	Carminative (prevents or eases abdominal flatulence)
Skullcap	Sedative (acts as a calming agent)
Tansy	Bitter (has a bitter taste, stimulate receptors in the tongue)
Vervain	Expectorant (loosens mucus in the trachea to aid elimination)

fresh, when in season. Some herbs are not very palatable, so may need to be disguised in food. Herbs act by exerting a direct medical or nutritional effect on the body and its processes.

It is vital in herbal medicine that plants should be identified correctly. They should be harvested from unpolluted areas, wherever possible, and should (if cultured) be grown without the use of modern agrochemicals. It is also advisable that, where possible, indigenous species should be used, because they may prove more suited to the patient's constitution than exotic herbs.

A trend in modern herbal medicine is the science of defining specific supposed 'active' ingredients, then extracting and purifying them and using them in isolation. This is not holistic medicine and carries inherent dangers, which do not attach to using whole plants. Ingredients of the whole plant tend to act in synergy and to balance each other in nature, whereas humans disturb this holistic balance with their 'interference'. Many products are now being marketed in this way. It is then but a small step to the alteration of these supposed active molecules and the marketing of modern patent drugs. Adverse side-effects have become an accepted part of modern medicine.

ACUPUNCTURE

Acupuncture is a part of Traditional Chinese Medicine (TCM). However, there are many modern adaptations, made in the light of experience and modern needs. It is used in both humans and animals, with dogs being no exception.

Because of its very specialized and complex application, it is recommended that acupuncture should be carried out only by a vet with the appropriate experience and expertise. In some countries the use of acupuncture in animals is legally restricted to veterinary application.

HOW ACUPUNCTURE WORKS

Typically, acupuncture consists of using needles to stimulate certain points in the body, to balance energy flow and quality. However, the stimulation of points may also be performed electrically (electro-acupuncture) and by means of laser, massage (acupressure), injection, implant and magnets.

TCM, which is around four thousand years old, teaches that the body energy (called Qi – pronounced 'chi') flows rhythmically around the body through meridians or channels, on a 24-hour cycle. There are 12 meridians on each side of the body, with the Qi flowing in each for about two hours, thus giving the 24-hour rhythm. Modern science has now confirmed what is called the 'circadian rhythm', thus lending support to this model.

Traditional Chinese Medicine also proposes the theory of yin and yang – the eternal opposites – of which Qi and every dynamic system are composed. Health depends upon the proper balance of yin and yang. Imbalance between yin and yang, or disruption of the rhythm or flow, gives rise to disease.

Needling (or other means of point stimulation) can restore the balance. However, spinal manipulation, internal medicine (usually Chinese herbs, but also

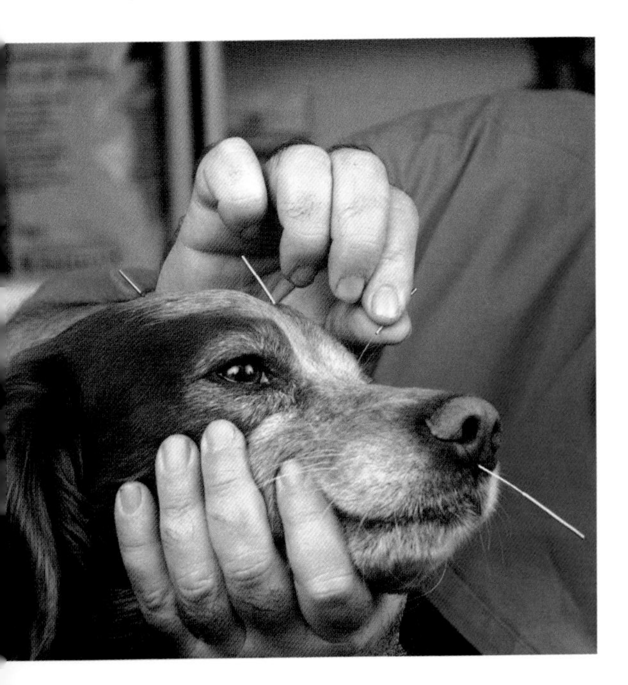

◀ *Dogs accept acupuncture with equanimity or appear really to enjoy it – relaxation is very common.*

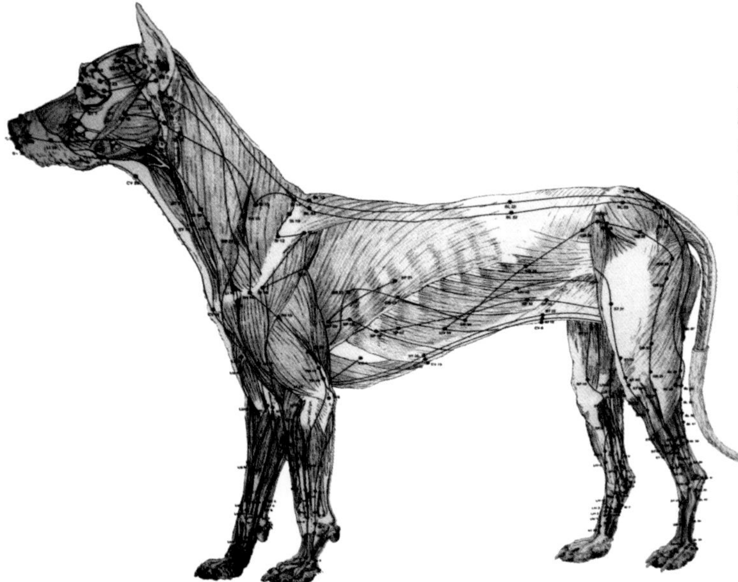

◀ *Qi flows through meridians or channels - the illustration gives one version of the position of the meridians.*

Western herbs or homeopathy), diet and lifestyle regulation are used in support of needling in the truly holistic practice of TCM. In the modern West we tend to use needling alone to try to achieve the medical benefits of acupuncture. The author believes that this rationale is incomplete and will result in more failures than the use of the fully holistic application of TCM.

While this model of life, health and disease is very different from that of modern Western medical science, using acupuncture according to this model is able to produce results.

ACUPUNCTURE FOR DOGS

Acupuncture is most frequently used on dogs for the treatment of locomotor problems, such as back and neck complaints, lameness, paralysis, arthritis and Chronic Degenerative Radiculomyelopathy (CDRM). Acupuncture is also employed as a means of pain control.

Those vets most experienced in this form of therapy will also use it to tackle conditions such as liver and digestive problems, uveitis (inflammation inside the eye), epilepsy and other chronic diseases.

ACUPUNCTURE AS HOLISTIC THERAPY

As a part of TCM, acupuncture – if practised in a traditional way, combined with manipulation, diet, lifestyle adjustment and internal medicine – is a deeply holistic form of therapy. It is used in veterinary medicine, and also as first-line medicine by a large proportion of the human population of the world.

See also *Herbs (48–49), Homeopathy (52–53), Chiropractic and Osteopathy (58–59), Locomotor Problems (70–71)*

HOMEOPATHY

Homeopathy is medicine practised according to the 'Law of Similars'. That is a natural phenomenon discovered formally by Dr Samuel Hahnemann (1755–1843) during the late 18th century. It has continued to this day, using principles derived from Hahnemann's extensive and meticulous writings, but with modern developments of the science.

Hahnemann lectured in Leipzig (*circa* 1813) on the use of homeopathy in animals. This practice has been extensively developed and put to practical use ever since, both for farm animals and in the home, for domestic pets.

HOW HOMEOPATHY WORKS

Homeopathic medicines are derived from plant, mineral and animal sources, diluted to extreme levels and applied according to the Law of Similars (that is, any substance that can provoke symptoms or signs in a healthy body can be used to stimulate cure of disease that displays similar signs or symptoms). Homeopathic pharmacies prepare the medicines with scrupulous attention to detail and hygiene, and with careful recording of the batches.

Medicines are selected not according to their pharmaceutical properties, but according to the symptom picture that the substance can provoke in a healthy body. They are extremely dilute, so cannot cause adverse or toxic effects, thereby enabling the use of many otherwise toxic substances. The dilutions are often so extreme as to reach sub-molecular levels, giving rise to some criticism of the method by the scientific community. The misunderstanding occurs because modern medical science tends to think in terms of the pharmacological action of a medicine rather than the biological (bio-energetic) stimulus and response.

▲ *Homeopathic medicines are safe on account of their extreme dilution – many such medicines are available over the counter.*

HOMEOPATHY FOR DOGS

Homeopathic medicines can be used in a simple and effective first-aid context or in serious acute or chronic disease. However, the deeper or more serious the disease, the more skill and homeopathic understanding are required. Many homeopathic medicines are available over the counter for domestic use, but for any serious disease veterinary advice should be sought. There is an expanding number of vets trained in the use of homeopathy.

Homeopathic medicines may be given as tablets, powders, alcoholic tinctures (not suitable for dogs) or as aqueous solutions. They are usually highly acceptable to animals and can be put in the mouth, rather than forced down the throat. Many dogs will readily take them from a saucer or dropper.

Homeopathy treats the animal as an 'energetic whole', not as a collection of symptoms or signs with a specific 'scientific' disease name. Because the animal itself, not the named disease, is being treated, vets need a great deal of information about the patient and his medical history, background and home environment. When you visit a homeopathic vet, this entails answering seemingly strange questions, which often appear quite unrelated to the specific problem for which the dog is being presented. The holistic vet needs to identify and remove those factors in the life of the patient that may impede healing, in order to maximize the chances of a cure.

One specialized area of homeopathic medicine is the prevention of infective disease by the use of nosodes (see page 35). These are medicines made from disease material and are used – a little similar to the principle of vaccination, but without the risks – for the prevention of specific diseases. This method is without currently accepted proof of efficacy and should only be used under expert veterinary guidance.

See also *Vaccination and Alternatives (34–37), First-aid Conditions (64–65)*

▼ Arnica montana *is used for the treatment of any injury or trauma.*

AROMATHERAPY

The use of 'essential oils' derived from different plants is described as aromatherapy. However, not all the medicinal agents used under this umbrella heading are oils. The process by which they are derived is distillation, so they include all the aromatic and volatile compounds of plants, and not just oils.

HOW AROMATHERAPY WORKS

Because these medicines are derived from plants, there is a relationship to herbal medicine, but aromatherapy is a very specialized branch of herbal medicine and the compounds and oils can be extremely powerful – a little can achieve much. It is therefore strongly advised that their general use in animals is confined to expert vets. However, some of the medicines do lend themselves to home use in animals.

The route of absorption of the medicines is via the olfactory (sense of smell) receptors, thus achieving very rapid transfer to the brain and the bloodstream and distribution around the body as a whole.

AROMATHERAPY FOR DOGS

Dogs are generally very willing to accept this form of medication, and their cooperation is encouraged by the fact that medicine is not forced into their mouths.

◀ *Lavender is a commonly used relaxant and calmative – animals respond very well to its calming properties.*

Aromatherapy can be administered by steam inhalation, with the dog's head being carefully placed near a bowl of steaming water into which a few drops of the oil or compound have been placed. Alternatively, oils can be burned (taking care with flame) in specialist oil burners or as candles, or neat essential oils can be sprinkled in the room; they can also be applied to furniture or bedding. An open bottle may even be held near the dog's nose. A historic example of medication by olfaction is sal volatile (smelling salts or ammonium carbonate) for revival after a faint.

Just as in herbal medicine, the capabilities of aromatherapy medicines extend to deep and serious medical use, for a wide range of chronic and acute illnesses in dogs, whether serious or non-serious. However, home users are not recommended to attempt to deal with serious illness in this way.

Aromatherapy compounds also have the ability to deter insects, such as fleas and lice. Tea tree, neem, garlic, lemongrass, lemon, eucalyptus, cedar and pennyroyal are known insect repellents, but the last is also known as an abortifacient (induces abortion in animals and humans), thereby reinforcing the message that wholesale and uneducated use of aromatherapy is not without danger. With the exception of pennyroyal, a few drops of the above oils can be diluted in a bowl of water, and the water can then be combed through the coat. The precise mixture of oils can be varied from day to day. Sadly, this measure is of little use in protecting against ticks.

See also *Fleas (38–39), Lice, Mites and Ticks (42–43), Herbs (48–49)*

COMMON AROMATHERAPY MEDICINES

Essential oil	Action
Basil	Digestive
Bergamot	Analgesic (eases pain)
Camphor	Stimulant
Chamomile	Nervine (relaxes and eases anxiety)
Clove	Anaesthetic (useful for painful or itchy lesions)
Eucalyptus	Expectorant (induces coughing) and decongestant
Fennel	Galactagogue (stimulates mother's milk)
Garlic	Disinfectant (useful internally or externally)
Lavender	Calmative (eases fears and anxiety)
Myrrh	Astringent and aids oral hygiene and health
Peppermint	Carminative (helps in cases of flatulent colic)
Rosemary	Stimulant and disinfectant (a useful additive for floor or surface cleaning)
Tea tree	Disinfectant and insect repellent

BACH FLOWERS AND TISSUE SALTS

Bach Flowers were the brainchild of Edward Bach (1886–1936), so they are a relatively modern branch of natural medicine. Bach was a homeopathic physician of importance, but his work and personal development eventually led him to a more intuitive approach, employing medicines that would access disease via its emotional roots.

He developed 38 remedies that are called, collectively, Bach Flowers (although, confusingly, they are not in fact all made from flowers). Worldwide, other flower essences have since become widely used – for example, Bush Essences from Australia.

For the most part, the remedies are made by first floating flowers in pure water, in direct sunlight, for three hours. Woodier parent material, or flowers that appear in periods of reduced sunlight, are prepared by boiling them in water for 90 minutes. The resultant solutions are mixed with brandy to make the mother tinctures.

BACH FLOWER REMEDIES FOR DOGS

The remedies are gentle and effective, giving no side-effects, and are well accepted by dogs. They are suited to the intuitive nature of animals, being used on the basis of the emotions, demeanour and mood of the patient, and treating even organic disease via the mind and emotions. The only real difficulty is in determining the relevant emotion or mental state of the patient. Physical signs are not listed among the indications. The remedies are fully compatible with homeopathic treatments, provided they are properly integrated with them.

COMMON BACH FLOWER REMEDIES

Remedy	Problem
Agrimony	Stoical suffering, mental pain with a calm demeanour
Cerato	Lack of confidence
Chicory	Possessiveness and jealousy
Impatiens	Impatience, a hurried demeanour
Mimulus	Shyness, a reserved nature
Rock rose	Terror
Vervain	Tension and anxiety
Walnut	Sensitivity to change
Rescue Remedy	Shock, distress or panic

COMMON TISSUE SALT MEDICINES

Tissue salt	Problem
Calc. fluor.	Supports tissue elasticity in the ligaments and blood vessels and helps the bones and teeth
Calc. sulph.	Its cleansing action helps to maintain a healthy skin and treat furunculosis (an acute skin disease), pimples or acne
Ferrum phos.	For respiratory problems, blood health, sore throats and bleeding, and to help aged dogs
Kali phos.	Used in cases of nervous tension and nervous exhaustion
Kali sulph.	For catarrh and skin ailments
Natrum mur.	For fluent coryza (catarrhal inflammation) and loss of smell or taste
Silica	Aids the structure of the skin and nails and is used to reject purulent infection

TISSUE SALTS

In 1880 Dr Wilhelm Schuessler (1821–98) published his system of the 12 so-called Biochemic Tissue Salts. He postulated that these 12 salts were all that was required to restore health to the body, since any imbalance of them in the cells would give rise to disease.

TISSUE SALTS FOR DOGS

Tissue Salt medicines are prepared by the same dilution and succussion method as homeopathic medicines, to a 6x (d6) potency. The potentization process is the method by which a material is serially diluted, to produce an 'energy medicine', for homeopathic and Tissue Salt medicines.

Typically, a 'mother tincture' is prepared and one drop is diluted with nine drops (on the decimal scale, written 'd' or 'x') or 99 drops (on the centesimal scale, written 'c' or 'CH') of alcohol/water solvent and succussed (shaken). This process is repeated until the desired dilution or potency has been reached (for example 30c, d6).

Tissue Salts are usually obtained as friable soft tablets and are well accepted by dogs. The author has used these medicines (and a new range that he has personally developed, in the light of modern agriculture, diet and lifestyle developments) not only in the treatment of ailments and illnesses, but also to help to correct deficiencies of minerals that have occurred through problems of assimilation or absorption.

While seemingly very simplistic in approach, these medicines – when used according to the stated indications – have resulted in positive outcomes.

See also *Homeopathy (52–53)*

CHIROPRACTIC AND OSTEOPATHY

Chiropractic and osteopathic manipulation are two branches of manipulative therapy, which act in support of medical intervention. Within each category there are widely differing techniques, so a complete definition of either technique is not possible.

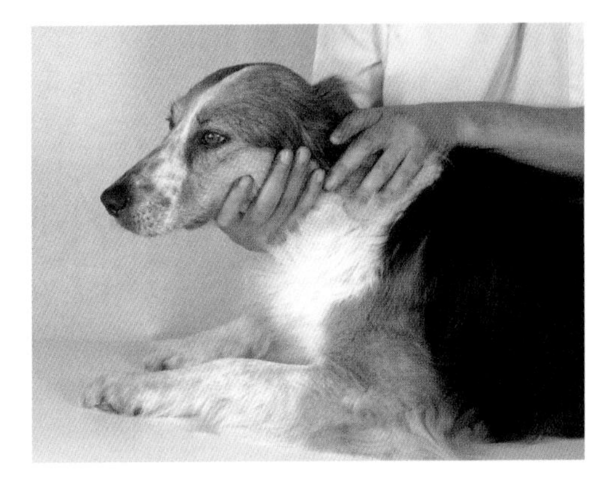

▲ *Chiropractic manipulation is a gentle means of restoring both correct somatic function and normal alignment to the musculoskeletal system.*

HOW MANIPULATIVE THERAPY WORKS

The skeletal system is composed of the axial skeleton and the appendicular skeleton. The head, neck, thoracic spine, lumbar spine, sacrum and tail form the axial skeleton and the pelvis operates as part of this system too. The appendicular skeleton consists of the limbs. The units of the skeleton (the bones) articulate with each other at joints, and muscles provide the motive power to operate the joints. Nerves emanate from the axial skeleton, to innervate (supply with nerves) the muscles. The correct integration of this system results in the fluid and pain-free motion that we expect from a healthy animal.

As soon as there is misalignment or muscle spasm, pain and incorrect function ensue, with possible nerve involvement and further functional alteration. The body compensates for this by incorrect posture and incorrect movement, often resulting in further damage as a result of the abnormal load being put on other parts of the musculoskeletal system. The prospect of a harmful downward spiral is then obvious. By gentle, judicious manipulation, muscle spasm can be relieved, joint alignment restored, nerve impingement released and correct posture and function restored. This is a reversal of the harmful spiral and could be termed a 'virtuous spiral'. This is the work of manipulative therapy.

MANIPULATION FOR DOGS

Canine conditions that are most clearly and directly helped by this intervention are neck pain, back pain, prolapsed disc, facial and cranial misalignment, pelvic misalignment and many cases of hind-limb weakness or paralysis. However, it may be surprising to learn that a large percentage of canine patients, on routine examination, display back, neck and pelvic problems that have often gone undetected and unsuspected.

Patients that have been successfully treated show obvious relief that may be

COMPLEMENTARY THERAPIES

It is interesting to note that even in Traditional Chinese Medicine, devised so many millennia ago, chiropractic-type manipulation was used in support of acupuncture in order to obtain better and longer-lasting results. The modern application of medicine (whatever system is preferred) would do well to integrate the various therapies for optimal results, rather than restricting a patient to one or other therapy or system exclusively.

almost immediate. In performance and working animals, the benefit shows not just in patient welfare, but also in the dogs' enhanced performance.

These manipulative therapies are not stand-alone treatments, however. They operate in support of acupuncture, herbal medicine, homeopathy and conventional medication. In many cases they add an essential extra dimension, without which success cannot be achieved. Depending upon how long the problem has persisted before treatment, muscle tone and strength and joint flexibility may also require the use of physiotherapy to complete the restoration of normal and pain-free function.

THE CANINE SKELETON

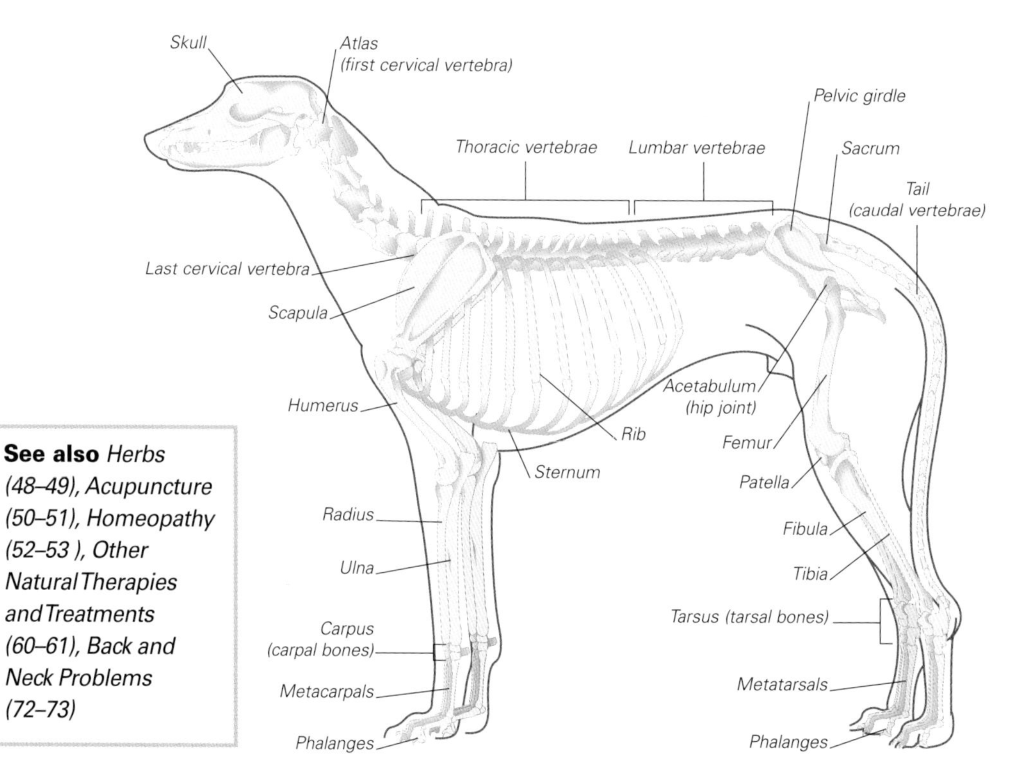

See also *Herbs (48–49), Acupuncture (50–51), Homeopathy (52–53), Other Natural Therapies and Treatments (60–61), Back and Neck Problems (72–73)*

OTHER NATURAL THERAPIES AND TREATMENTS

Here some of the best-known physical therapies are discussed. These therapies act in support of medical systems such as herbal medicine, homeopathy, acupuncture and modern medicine, and can also support chiropractic or osteopathic manipulation.

PHYSIOTHERAPY

This is a non-specific umbrella term, used to describe many forms of hands-on and exercise therapies that work to restore normal function and range of movement to the musculoskeletal system. It can also cover the use of instruments such as LASER, ultrasound, TENS and faradic stimulation. It is not a system of medicine (alternative or conventional) in its own right, but should be used as an adjunct to medical input.

MAGNET THERAPY

This therapy uses magnets applied to specific areas of the body, to encourage circulation and stimulate healing. Magnets are usually applied for limited periods of each day, rather than being used for 24 hours. Small magnets have been designed to be applied with adhesive tape, and there are also blankets containing pouches in which magnets can be inserted in the desired position.

MASSAGE

A hands-on therapy that anyone can learn, massage releases tension in the muscles and enables the musculoskeletal system to operate in a more normal manner, with resulting benefits in posture and motion. It can be quite strenuous for the masseur or masseuse at first, but becomes easier with practice. It is very rewarding, since a dog will clearly demonstrate where it enjoys or needs a massage and will find a way to indicate whether more or less pressure is required,

▼ *LASER therapy stimulates the healing of skin, tendons, ligaments and muscles.*

and when sufficient massage has been received in one place or in one session.

LASER THERAPY

LASER (Light Amplification by Stimulated Emission of Radiation) therapy employs cohesive light – usually infrared, but other colours may be employed for specialist purposes – to stimulate the healing of tissues such as muscle, tendons, ligaments, fascia, tendon sheaths and joint capsules. The probe contains a light-emitting diode, which can be set to deliver the beam at different pulsed frequencies. Different probes are required for different wavelengths (colours). LASER therapy differs from LASER surgery in that it uses what is called a 'Cold LASER', which cannot destroy tissues. This makes it a very safe form of therapy, but the beam can damage the retina of the eye, if it is accidentally aimed or reflected towards the pupil.

ULTRASOUND THERAPY

This employs the vibrations of ultra-high-frequency sound transmitted to areas of the body by means of a probe or head. This creates a powerful and localized form of massage and stimulates healing in soft tissues such as the muscles and tendons. The dog's coat is usually shaved and a gel applied, to achieve good contact with the skin. Care has to be exercised not to apply ultrasound over bone, and not to any area for too long. It is advisable that an appropriately skilled vet should be consulted.

TELLINGTON TOUCH

Tellington Touch is a method of improving balance and coordination, via hands-on contact and exercises. It works on the principle that a dog can develop patterns of tension, which may create stress elsewhere in the body and affect the dog's general well-being and behaviour. Signs of tension areas in a dog may include a local change in fur pattern, a patch of dry skin or exaggerated skin reflexes when touched. It is thought that the techniques may activate neuromuscular pathways, helping the body to achieve improved posture and movement for itself.

▲ *Hydrotherapy is a rapidly developing field, which can provide valuable support for treatment of musculo-skeletal problems.*

See also *Massage (14–15), Herbs (48–49), Acupuncture (50–51), Homeopathy (52–53), Chiropractic and Osteopathy (58–59)*

HEALTHCARE AND YOUR VET

While you are ultimately responsible for the health of your dog, advice and support from different sources can be essential to ensure that any health and management programme is the best that can be devised. Your vet is trained in the recognition of early signs of disease, and knows the range of ailments and problems that can affect your dog's health.

Regular veterinary check-ups are recommended. These should take place at least annually for younger dogs and at least twice yearly in the case of older dogs. If the check-up is with a holistic vet, he or she will be fully conversant with any holistic measures that you have in place. If not, some discussion and explanation will be necessary. If there is no holistic vet in the locality, it can be arranged that a more distant holistic vet communicates with your local vet. Even if no holistic vet is involved, regular veterinary support is still recommended. Your vet will be in touch with the latest developments and will keep up to date with the prevailing health situation.

An annual blood test is a good precaution, in support of a clinical examination. You may also consider it worthwhile regularly taking in urine and faeces samples, in order to monitor your dog's health. If any sign of disease is found at a regular check-up, your vet may recommend investigations aimed at positively

◀ *Your vet is trained in the recognition of early signs of disease – an annual or six-monthly check-up is a good precaution.*

identifying the problem, so that a relevant treatment programme can be put in place.

WORKING WITH YOUR VET

If your wish is for one or other form of holistic or natural veterinary treatment to be employed and your local vet is not a holistic vet, then you may need to travel further to obtain that help. However, it is still important to keep your local vet closely in touch with whatever treatments are being employed and with your dog's progress, so that any potential emergency response can be appropriate to the whole case. If more than one vet is working on behalf of your dog's health, it is vital that they are in communication and cooperating with each other.

If you have the confidence or experience to carry out some home treatments – depending upon what is wrong with your dog – then again it is advisable to keep your vet in touch with what you are doing and how things are progressing. It may even be that you can work with your vet, using your treatments and his or her diagnostic and monitoring skills.

Whatever arrangement is appropriate to your individual circumstances and your dog's special needs, it is never constructive for there to be arguments and conflict about the form of treatment being used.

Communication is important in maintaining cooperation, and discussion in advance of any problems is valuable. Your wish will be for your dog's optimum health and well-being at all times, and your vet's motivation will be the same, thus ensuring that you work together for your dog's welfare.

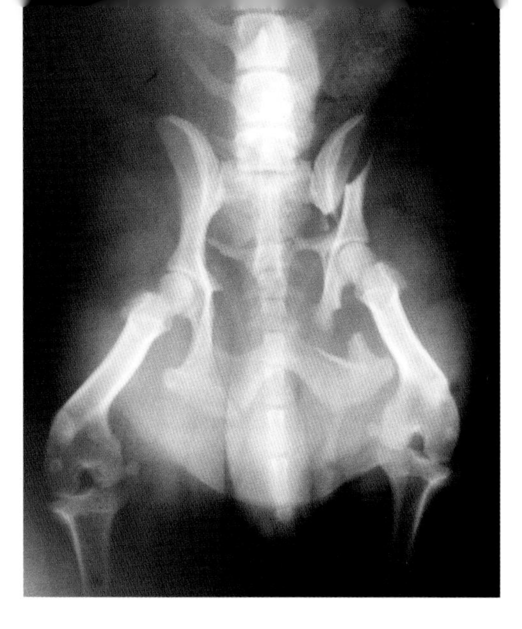

▲ *A veterinary clinic has diagnostic equipment, such as X-ray machines, along with surgical and care facilities, in case investigations or specialist care are needed.*

REMEMBER...

Whether you use vaccination or natural alternatives for your dog, a regular veterinary check-up is advisable, in order to ensure that your health programme is on course and that there are no signs of ill health. If you have a holistic health programme for your dog, then you probably already use a holistic veterinary clinic or at least have a holistic vet on your veterinary team.

See also *Vaccination and Alternatives (34–37)*

FIRST-AID CONDITIONS

CAUTION

This section of the book (pages 64–81) is designed to act as a guide to common ailments and available remedies, not as a home treatment manual. You are always advised to consult your vet when in doubt, or if simple measures fail to bring relief.

In emergency situations or when injury occurs, prompt action can make all the difference. The following homeopathic and other natural medicines can be given with safety while awaiting attention from your vet, without fear of interfering with any other treatment that may be required.

Various therapies are mentioned, to suit different people's first-aid cupboards. Homeopathic medicines should be given internally in a 6c or 30c potency. Homeopathic and aromatherapy remedies can be given even to a collapsed animal, since swallowing is not required.

A GUIDE TO AVAILABLE THERAPIES AND REMEDIES

Problem	Therapy or remedy
Abrasions	Hypericum and Calendula lotion topically; Hyssop (*aro*)
Abscess	Hepar sulphuris (*hom*) if acute, Silicea (*hom*) if chronic; warm saline bathing
Anaemia	Ferrum phos. (*t/s*); Cinchona (*hom*) – give both and feed red/purple vegetables, black olives, liver
Bites (dog or cat)	Collidal silver and/or Hypericum and Calendula lotion topically; Hepar sulphuris (*hom*), Arnica (*hom*) – give both; Hyssop (*aro*)
Bites (insect)*	Hypericum and Calendula lotion topically; Aloe-vera juice (*herb*) topically; Apis (*hom*)
Bites (snake)	Lachesis (*hom*) – the Bushmaster, Vipera (*hom*) – the Adder, Crotalus (*hom*) – the Rattler: professional advice may be needed to decide which to use
Bleeding	Hamamelis (*hom*) – dark oozing, Aconitum (*hom*) – red gushing, Millefolium (*hom*) – red seeping, Phosphorus (*hom*) – bleeding tendency
Bloating	Nux vomica (*hom*), Carbo vegetabilis (*hom*); Peppermint (*aro*) – all can be given concurrently

Problem	Therapy or remedy
Bruises	Arnica (*hom*); Comfrey (*herb*) – both can be given topically and internally; Bergamot (*aro*)
Burns	Urtica (*hom*), Arnica (*hom*), Cantharis (*hom*) – give all three
Collapse (*see opposite*)	Carbo vegetabilis (*hom*), Aconitum (*hom*) – give both; Camphor (*aro*), Rosemary (*aro*)
Convalescence	Cinchona (*hom*), Phosphoric acid (*hom*) – give both
Crushing injury	Arnica (*hom*); Hypericum (*hom*) – give both
Cuts	Hypericum and Calendula lotion topically; Staphisagria (*hom*)
Foreign body	Myristica (*hom*), Silicea (*hom*); poulticing (for example, Magnesium sulphate/Glycerine paste, also known as Morrison's Paste)
Fractures	Arnica (*hom*), Symphytum (*hom*); Comfrey (*herb*), Boneset (*herb*) – all can be given concurrently
Glands (swollen lymph nodes)	if generalized, this could be a general illness or a cancerous condition (see pages 88–89); if localized, it is probably the result of a local source of infection. Hepar sulphuris (*hom*) would be appropriate for the latter
Haemorrhage	see *Bleeding*
Heatstroke	Glonoinium (*hom*), Aconitum (*hom*) – give both
Paralysis/disc prolapse	Aconitum (*hom*) – pain and shock, Arnica (*hom*) – pain and healing, Apis (*hom*) – spinal-cord oedema, Nux vomica (*hom*) – spasm, pain and problems urinating or defecating, Hypericum (*hom*) – nerve damage and pain; Bergamot (*aro*) – pain. All may be useful and can be given concurrently
Poisoning	Nux vomica (*hom*) and seek professional advice for the specific poison
Post-operative recovery	Arnica (*hom*) – injury, Staphisagria (*hom*) – surgical injury, Hypericum (*hom*) – pain, Nux vomica (*hom*) – anaesthetic after-effects
Puncture wounds	Ledum (*hom*); Hypericum and Calendula lotion topically; saline bathing
Scalds	see *Burns*, plus Apis (*hom*)
Shock	Aconitum (*hom*); Lavender (*aro*)
Stings (insect)*	Apis (*hom*); Aloe-vera juice (*herb*) topically
Stings (nettles)	Urtica (*hom*); Dock-leaf juice (*herb*) topically

Abbreviations

aro aromatherapy
b/f Bach Flowers
herb herbal
hom homeopathic
t/s Tissue Salts

* For applying topically to insect stings and bites, vinegar or baking soda can be very effective. For wasp stings, vinegar helps; for bee stings, baking soda helps (and removes the sting). For other stings and bites, some trial and error may be needed to establish which is most helpful.

See also *Herbs (48–49), Acupuncture (50–51), Homeopathy (52–53), Other Natural Therapies and Treatments (60–61), Back and Neck Problems (72–73)*

GASTROINTESTINAL PROBLEMS

Gastrointestinal problems are often observed by vomiting or by changes in faeces or defecation. They may also be noticed because of flatulence, changes in appetite or weight loss. These signs can be transitory or may indicate deeper and more serious disease, so veterinary help is recommended.

ANAL GLANDS

If anal glands keep filling, homeopathic Silicea may help. It is wise to look into the diet. Is there enough roughage (vegetable fibre)? Herbal Psyllium husks may help to bulk out the stool. The anal glands serve an excretory function, and unsuitable dietary ingredients (chemicals and toxins) can cause problems.

APPETITE

A decreased or increased appetite may signal deeper health problems. Ask your vet for a check-up. Homeopathic Lycopodium or Nux vomica may help a poor appetite; Calcarea phosphorica or Phosphorus may help a depraved appetite; and Sulphur may help with coprophagy (eating faeces).

BLOATING

Homeopathic Nux vomica and Carbo vegetabilis are emergency remedies for this condition. Herbal Peppermint or aromatherapy Peppermint oil may also be valuable. *This is a veterinary emergency.*

CONSTIPATION

Check with your vet for possible causes. Homeopathic Sulphur, Silicea, Nux vomica or Alumina may help. Herbal Psyllium in the food husks forms a lubricating gel and bulk aperient to help bowel movement. Herbal Prune juice is a known laxative. In the short term, herbal Senna may be helpful. Feed the dog plenty of vegetables.

DIARRHOEA

Acute diarrhoea may be necessary to eliminate toxins from unsuitable food. Consult your vet if it persists. If it is caused by gastroenteritis, then homeopathic

◀ *If there is straining to produce diarrhoea, this is an important clue.*

Arsenicum (if there is a dry mouth), Mercurius solubilis (for a wet mouth) or Mercurius corrosivus (for a wet mouth and dramatic straining) may be helpful.

FLATULENCE

If this is chronic, the diet may be unsuitable for the individual dog or there may be a veterinary problem. If it is a temporary or occasional problem, then homeopathic Lycopodium and Carbo vegetabilis will usually help, as will herbal Charcoal biscuits or Peppermint or Peppermint oil (aromatherapy).

INCONTINENCE

If a dog involuntarily defecates in the house where it is lying or on the move, this is faecal incontinence and may indicate a medical problem. If it is an 'old age' problem, homeopathic Causticum may help.

LIVER PROBLEMS

There are many remedies in natural medicine for liver complaints. Homeopathic Lycopodium (flatulence), Nux vomica (constipation), Chelidonium (jaundice, especially obstructive), Phosphorus (inflammatory liver with jaundice) and Carduus (swollen liver) are possibly the most-used remedies. Herbal Milk Thistle is also very well known in this connection.

PANCREATIC INSUFFICIENCY

This condition may be seen in German Shepherd Dogs, among others. Supplementation with enzymes may be necessary (your vet will supply these), but homeopathic Lycopodium, Iodum, Phosphorus or Phosphoric acid may help.

▲ *Oral health (especially gums and teeth) is vital to overall health and can be affected by diet.*

TEETH AND GUM DISEASE

Sound gum and dental health is essential for your dog's general well-being, and a diet containing regular chunky raw meat and fresh knuckle bones to chew (see page 20) is important. If there is gingivitis, Myrrh (as a herbal tincture or aromatherapy oil) is an excellent local treatment. Homeopathic Mercurius solubilis (for profuse saliva) or Pyrogen (for rotten odour) is also helpful. Tartar should be removed using the thumbnail, or by your vet.

TRAVEL SICKNESS

Homeopathic Petroleum, Tabacum or Cocculus usually helps. Aromatherapy (Lavender) may also prove useful, for calming the dog.

VOMITING

Vomiting can be a sign of serious disease, so consult your vet. If gastroenteritis is diagnosed, Arsenicum (dry mouth), Mercurius solubilis (wet mouth, yellow frothy vomit) or Mercurius corrosivus (wet mouth, clear mucoid vomit) can put a rapid stop to it.

SKIN PROBLEMS

Skin disorders can be complex and often betray an underlying constitutional problem. Most homeopathic and herbal medicines have some action on the skin, making a remedy selection quite complex. Diet has a large influence on skin health.

ALOPECIA

Loss of hair can be a sign of deeper disease (consult your vet). It may be the result of hormonal imbalance or can occur as a result of mange. If there appears not to be another cause, then homeopathic Sepia or Arsenicum may be useful.

ANAL FURUNCULOSIS

This painful condition is seen mostly in the German Shepherd breed. Surgical removal of the anal glands may be necessary, if medical efforts fail, but LASER therapy, homeopathic Silicea or Myristica and topical Boric Acid powder are remedies that are commonly tried.

▼ *Excessive itching may be a sign of allergy or ear problems.*

COAT-SHEDDING/EXCESSIVE MOULTING

This may be caused by central heating or it may be of dietary or hormonal origin.

CYSTS

Sebaceous cysts are usually non-serious, but they can become infected. Homeopathic Thuja, Graphites or Silicea may help (or Hepar sulphuris, if the cysts become painful and infected).

DANDRUFF/SCURF

Scurf may be due to general dietary or skin problems. Feeding omega-6 and omega-9 oils can help to prevent the problem. Homeopathic Arsenicum, Pulsatilla or Sulphur may be suitable, according to constitutional indicators.

ECZEMA

Eczema is a loose term that usually applies to skin disorders characterized by dry patches. There may also be itchiness or some seborrhoea (greasiness). Useful medicines are homeopathic Sulphur, if the dog likes cool conditions and is made worse by warmth; Graphites, if the dog likes warmth, but is made worse by warmth and tends to be overweight and lazy; and Psorinum, for a dog that likes and is improved by warmth. Zinc and Castor-oil cream or ointment may also help.

FURUNCULOSIS

Furunculosis (recurring boils) is a sign of immune imbalance, permitting *Staphylococcal* infection. Homeopathic Graphites or Sulphur may help, along with ensuring that the dog's diet and general health are optimal.

INTERDIGITAL CYSTS

This is a painful septic-type condition of the interdigital spaces between the dog's toes. Bathing with a warm, strong saline can soothe the complaint and draw the pus. Homeopathic Silicea or Myristica are among medicines that may help.

ITCHING/PRURITUS

Excessive itching – if not caused by fleas, lice or mange – may be dietary or immune-related. Seek veterinary advice. If it is allergic or constitutional, then homeopathic constitutional prescribing can usually help. Common medicines used are: Agaricus, Calcarea carbonica, Graphites, Psorinum, Pulsatilla, Ranunculus and Sulphur, prescribed according to the constitutional indicators.

NAILS

Breaking, splitting or very soft nails are a sign of disease, which may be a local infection or part of more general skin problems or autoimmunity. Sometimes the nailbed can become infected. Dietary zinc, sulphur and biotin must be adequate. Homeopathic Silica or Graphites may be relevant.

RINGWORM

Ringworm is a zoonosis (a disease of animals that is infectious to humans) and a fungal infection. It may or may not be itchy. Hair loss and greying skin may appear. Ultraviolet light is one of the tests that can be used. A laboratory can also examine a skin scraping. Ringworm is usually rapidly resolved with homeopathic treatment (e.g. Bacillinum).

TUMOURS AND WARTS

Warts may be helped by homeopathic Thuja, Calcarea carbonica, Lycopodium or Causticum. Herbally, local application of the sap of Greater celandine, Dandelion or Caper spurge may remove the wart. For tumours, see pages 88–89.

WET ECZEMA/HOT SPOTS

Seek help from a holistic veterinarian.

See also *Fleas (38–39), Lice, Mites and Ticks (42–43), Gastrointestinal Problems (66–67), Allergy and Atopy (84–85), Autoimmune and Endocrine Disorders (86–87)*

LOCOMOTOR PROBLEMS

Diet can have a very powerful influence on skeletal health, in both the long and short term. Chiropractic evaluation is important in cases of locomotor problems to ensure correct skeletal alignment, posture and optimal weight distribution. Massage and/or other physical therapy can be very beneficial for musculature, if it is injured or in spasm.

ARTHRITIS

This term literally means inflammation of a joint. It is called osteoarthritis if there are bony changes around the joint. A simple sprain can also be 'arthritis'. Rarely there may be septic (infected) arthritis. Common helpful homeopathic medicines are Rhus toxicodendron (worse for cold and damp and stiff on rising, but limbers up), Bryonia (worse for any movement), Ruta (sprains and strains), Calcarea fluorica (osteoarthritis) and Rhododendron (worse in thundery weather). Herbally, common medicines are Willow bark, Meadowsweet, Comfrey and Devil's claw. Acupuncture, acupuncture-by-LASER and LASER Therapy can also help, arthritis being some of the most common fields of application of these therapies in dogs.

CRUCIATE-LIGAMENT INJURIES

Boisterous dogs can damage or tear the anterior cruciate ligament in the stifle joint (knee). Surgery is commonly offered, but an integrated programme that includes LASER therapy, acupuncture, chiropractic manipulation and homeopathy can help some patients to avoid surgery. Homeopathic Ruta and/or herbal Comfrey are valuable supportive treatments.

▼ *For fractures, veterinary attention is essential and some form of fixation may be required.*

DEGENERATIVE JOINT DISEASE (DJD)
DJD is a chronic disease process. It can be eased by homeopathy (Ruta, Calcarea fluorica, Hekla lava) and herbs (Comfrey, Willow bark, Meadowsweet or Devil's claw). Acupuncture and LASER treatment may help.

FRACTURE
This is a break or crack in a bone. Herbal Comfrey or Boneset and homeopathic Calcarea fluorica, Ruta, Symphytum and Arnica are commonly used, along with external or internal fixation as necessary.

LAMENESS
Lameness arises from any pain in a limb or in the musculature of the shoulder or hip.
It can arise from nerve impingement in the neck or lumbar spine. Holistic veterinary advice should be sought before chiropractic treatment is undertaken and if the chiropractor is not a vet, the work should be done under vetinary supervision. Acupuncture or other natural treatments can help.

OSTEOCHONDRITIS DISSECANS (OCD)
OCD is a disturbance in the cartilage surface of the joint. Surgery is often offered, but dogs may respond to medical input as a first option. Acupuncture, homeopathy (Ruta, Calcarea fluorica, Caulophyllum, Ledum or Kalmia, according to the patient), herbs (Comfrey, Meadowsweet, Cleavers) and LASER therapy can be very helpful.

OSTEOSARCOMA
This is usually a very painful and aggressive cancer of the bone. For further information, see pages 88–89.

▲ *Meadowsweet (among other herbs) contains a salicylic-acid-like compound, with pain-killing and anti-inflammatory benefits.*

RHEUMATISM
This painful muscular disease can be helped in a similar way to arthritis (see above), through homeopathy, herbal treatments or acupuncture.

SPRAIN/JOINT INJURY
This is a form of arthritis (see above) and involves acute injury to the ligaments and joint capsule. LASER therapy and homeopathic Ruta are usually the treatments of choice in natural medicine. Support bandaging will help, if this is practical. If the joint is very swollen, homeopathic Apis and herbal Dandelion can help to bring down the swelling.

See also *Back and Neck Problems (72–73)*

BACK AND NECK PROBLEMS

As with locomotor problems (see pages 70–71), chiropractic evaluation of back and neck problems is vital, to ensure correct skeletal alignment, posture and optimal weight distribution. In cases of injury or spasm, massage and/or other physical therapy can be very helpful. Diet also plays an important role in skeletal health, both in the short term and the longer term.

CHRONIC DEGENERATIVE RADICULOMYELOPATHY (CDRM)

CDRM is a progressive degenerative condition of the spinal cord, seen mostly in German Shepherds, but also in Golden Retrievers and Rough Collies. A worrying number of cases have come on suddenly, after being given a double vaccine booster, because of lapsed vaccination programmes. Homeopathy and acupuncture have helped significantly in some cases. The outlook paradoxically appears better in dogs older than nine years at the date of onset.

FRACTURE

If a bone is fractured, herbal Comfrey or Boneset and homeopathic Calcarea fluorica, Ruta, Symphytum and Arnica are commonly used, along with external or internal fixation as necessary (see pages 70–71). In the case of spinal fractures, homeopathic Hypericum can be helpful in limiting nerve damage.

HEMIVERTEBRA

Hemivertebra, sometimes seen in Pugs and other brachycephalic breeds (those with short faces), is a malformation of one or more vertebrae, usually thoracic. This can impinge on the spinal cord, with resultant ataxia or paralysis and sometimes pain. While nothing can be done medically about the underlying structural problem, experience shows that the functional disturbance and pain can be helped in most cases. Natural medical input consists of an integrated programme of acupuncture, chiropractic manipulation, LASER therapy and homeopathy (Hypericum, Lathyrus and Plumbum are commonly used).

MISALIGNMENT

Any part of the skeleton can become misaligned, with resultant malfunction, pain and muscle spasm. Spinal misalignment is common in dogs and will impinge on nerve function, whether to the limbs, muscles or internal organs. Common areas of most stress are the top of the neck (atlanto-occipital joint), the lower neck (where the fore-limb nerves emerge), the thoracolumbar junction (a common site of disc prolapse, too) and the lumbosacral junction (where the hind-limb nerves emerge). Chiropractic manipulation is essential, supported by acupuncture, LASER therapy, physiotherapy and other modalities, as needed. Homeopathic Arnica, Ruta and Hypericum may also help with pain control and stimulus to healing (see pages 58–59).

PARALYSIS/DISC PROLAPSE

In dogs, particularly long-backed breeds such as Dachshunds, disc prolapse with subsequent pain and often paralysis is quite common. If the disc prolapses, it is usually very sudden and forces the pulp into the spinal canal, where it concusses the spinal cord. In the short term, oedema (swelling), inflammation and pain must be reduced. In the medium term, it is necessary to stimulate healing of the nerves; in the longer term, to prevent further discs becoming prolapsed. In cases treated with natural medicine without surgery, the results have usually been good. Acupuncture is the main treatment used, supported by very gentle chiropractic manipulation, LASER therapy and homeopathy. Steroids and surgical treatment are often offered in conventional practice.

▼ *Homeopathic Hypericum is used to help cases of nerve injury or impingement.*

SPONDYLOSIS

This is a degenerative condition of the joints of the spine, in which new bone forms around spinal joints (sometimes extensively). Affected dogs suffer pain, loss of spinal mobility, loss of spinal musculature and variable impairment of hind-limb strength and function. An integrated programme of natural medical input includes acupuncture, LASER therapy, magnet therapy, chiropractic manipulation and homeopathic Phosphorus, Hekla lava, Hypericum, Conium, Plumbum or Calcarea fluorica, depending upon the patient and the predominant signs.

WEAKNESS

Weakness from age-related degeneration should be distinguished from that of arthritis, CDRM or spondylosis. Age-related weakness can usually be helped by homeopathic Causticum or Conium.

WOBBLER (CERVICAL SPONDYLOPATHY)

This is a condition of the cervical spine resulting from imperfect development, possibly as a result of accelerated growth through over-adequate nutrition during the growing phase. It usually only occurs in larger breeds of dog. Integrated treatment with acupuncture, chiropractic manipulation and homeopathic Conium has often helped.

See also *Locomotor Problems (70–71)*

EYE AND EAR PROBLEMS

Ear problems are often part of a more general skin complaint and can be noticed by discharges, by ear scratching or rubbing, or by a change in the dog's ear or head carriage. Eye problems can be very frightening, but the eye often heals vigorously under homeopathic guidance and stimulus, sometimes aided by herbal treatment or acupuncture. Expert ophthalmological opinion may be necessary for correct diagnosis.

CATARACT

This is characterized by progressive opacity of the lens of the eye. It can be age-related or can come on as a result of eye injury. There are also congenital and hereditary cataracts. Homeopathic Calcarea fluorica and Silicea are often used; Calcarea carbonica, Phosphorus and Conium may help age-related cataracts. Senega may assist cataracts following surgery.

CONJUNCTIVITIS

Homeopathic Aconitum and Argentum nitricum are commonly used for inflammation of the eye membranes. Euphrasia can be used homeopathically or herbally and may be applied to the eye as a lotion.

CORNEAL ULCER

The most common homeopathic treatments are Conium and Mercurius corrosivus, both of which are relevant when there is great photophobia (sensitivity to light) and pain. Treatment has usually been very successful. Euphrasia eye lotion may also help, particularly if there is blueing. Several other medicines may be useful.

DRY EYE (KERATO-CONJUNCTIVITIS SICCA)

Some cases of dry eye may be helped to avoid surgery by using homeopathy (Senega, Veratrum album and Zincum metallicum have been used, depending upon the case).

ENTROPION

Homeopathic Borax is the remedy best known to treat this congenital condition, in

▼ *Eye discharges can be carefully and gently removed as shown, with a moist tissue.*

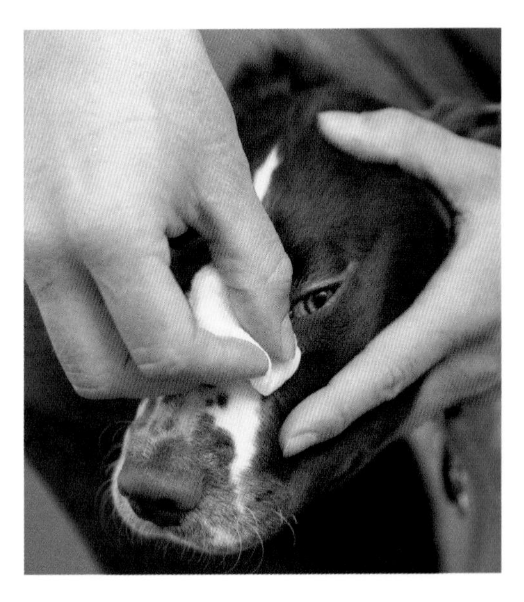

which the eyelid edges curl inwards. Surgery may be necessary.

EXTERNAL EAR INFLAMMATION (OTITIS EXTERNA)

This condition can usually be helped by a number of natural-medicine measures and management. Homeopathic Calcarea sulphurica, Causticum, Graphites, Kali bichromicum, Kali sulphuricum, Kreosotum, Mercurius solubilis, Psorinum, Pulsatilla, Pyrogenium and Sulphur are commonly used, according to constitutional and local indicators. Homeopathic Belladonna can help acute, painful inflammation. Management of this condition can be assisted by using aromatherapy (dilute Lavender, Rosemary and Tea tree, for example) or either Boric-acid powder or Iodoform powder.

GLAUCOMA

Glaucoma is swelling of the eyeball, usually with some pain. Homeopathic Belladonna, Gelsemium or Spigelia is often used in this condition, which is rarely seen before conventional drugs have been started, so it is difficult to know whether homeopathy alone can help.

HORNER'S SYNDROME

In this condition the eyeball is pulled back and the third eyelid comes across, the upper eyelid droops, the lower eyelid can be elevated and the pupil is constricted. It results from sympathetic damage, sometimes by head or neck injury. Acupuncture, supported by homeopathic Physostigminum, has helped some cases, but spontaneous recovery can also occur.

MIDDLE- AND INNER-EAR INFLAMMATION (OTITIS MEDIA AND OTITIS INTERNA)

These conditions usually result in head tilt with loss of balance. Homeopathic Causticum and Conium generally help to restore more normal function. There may be purulent infection, in which case homeopathic Hepar sulphuris can help. Aromatherapy (Lavender, Rosemary) can also be used.

PANNUS

The appearance of blood vessels on the cornea, with or without pigmentation, can usually be treated homeopathically, using Argentum nitricum, Aurum metallicum, Mercurius solubilis or Silicea, depending upon constitutional and local signs.

UVEITIS

This is a painful inflammation of the main internal layers of the eyeball. If not successfully treated, sight – and even the eye itself – can be lost. Experience has shown that integrated acupuncture and homeopathy (Belladonna, Gelsemium, Phosphorus or Spigelia, depending upon constitutional and local signs) can often save the sight.

VESTIBULAR SYNDROME ('CANINE STROKE')

This condition usually results in head tilt with loss of balance. Homeopathic Causticum and Conium generally help to restore more normal function. Aromatherapy (Lavender, Rosemary) can also be used, and acupuncture has assisted some cases.

Do not put any herbal or homeopathic remedy in the eye or ear unless directed to do so by an appropriately qualified veterinarian.

CHEST AND HEART PROBLEMS

Experience has shown that homeopathic medication, properly applied, can be very successful in bringing back good quality of life to many animals suffering from chronic heart disease. In numerous cases, conventional medication has proved unnecessary, and in others it has been possible to withdraw conventional medication. However, treating serious heart problems is not normally suitable for home medication.

HEART DISEASE

Homeopathic medication is extremely unlikely to interfere with any conventional drugs that your vet may prescribe, but you are recommended to seek the advice of an experienced holistic vet if you wish to use natural medicine, either alone or with conventional medicine. Herbal medicine, on the other hand, can conflict with conventional medicines or dangerously summate with it.

- Heart conditions that have been treated with good rates of success are: arrhythmia (abnormal heart rhythm), bradycardia (slow heart), cardiomegaly (enlarged heart), cardiomyopathy (disease of the heart muscle), heart failure (left, right or both), heart murmur (heart-valve problems) and tachycardia (fast heart).
- There are many homeopathic medicines that have an effect on conditions of the heart, and the following list draws attention to the most commonly used ones: Aconitum, Adonis, Apocynum, Arsenicum, Cactus, Carbo vegetabilis, Cinchona, Coffea, Convallaria, Crataegus, Digitalis, Gelsemium, Glonoinium, Kali nitricum, Lachesis, Laurocerasus, Lilium tigrinum, Lycopodium, Lycopus, Naja, Natrum muriaticum, Nux vomica, Phosphorus, Prunus, Spartium, Spigelia, Spongia, Strophanthus, Tabacum and Viscum album.

Readers with deep experience of homeopathy and access to homeopathic resources may wish to read textbooks on these medicines, to see which might be suitable for a particular patient, especially if the heart condition has not been successfully controlled by conventional medication.

ASCITES

If the right side of the heart is failing to pump blood sufficiently, fluid can accumulate in the abdomen. Herbal diuretics, such as Dandelion leaf, can help to clear the fluid, in addition to appropriate homeopathic heart medication (see above). Helpful homeopathic medicines are Adonis, Apis, Apocynum and Digitalis.

COUGH

There are many variations of coughs and many conditions that have it as a symptom or sign. Common homeopathic medicines that may help are: Aconitum (sudden onset, worse for cold), Antimonium tartaricum (rattling cough), Arsenicum (worse just after midnight), Bryonia (worse for movement, eating or drinking), Causticum (worse with warmth), Coccus (with fluid expectoration), Drosera (spasmodic cough, worse at night), Dulcamara (worse in damp weather), Ipecacuanha (with vomiting), Mercurius solubilis (worse in the hours of darkness), Phosphorus (worse in cold air), Pulsatilla (worse at morning and night, better in fresh air) and Spongia (if from heart disease). The essential oils Cedarwood, Eucalyptus or Hyssop may help coughs by reducing inflammation and softening mucus accumulation.

HEART COUGH

If the left side of the heart fails to pump blood properly, a 'heart cough' can result, owing to congestion of the lungs. Homeopathic Bryonia, Digitalis or Spongia may be helpful, along with herbal diuretic support, such as Dandelion leaf.

KENNEL COUGH

This can be a troublesome and disturbing condition, but is rarely serious. Even with antibiotic treatment, it can persist for weeks. If the correct homeopathic medicine is given, experience shows that kennel cough can be resolved in about three days. Homeopathic medicines that have been used are Aconitum, Coccus, Drosera, Ipecacuanha, Mercurius solubilis and Phosphorus, prescribed according to the individual patient and the signs shown (see Cough, above).

▶ *Excessive panting or shortness of breath can indicate heart or lung problems.*

PULMONARY OEDEMA/PULMONARY CONGESTION

If there is shortness of breath and a craving for fresh air, then homeopathic Apis and Carbo vegetabilis may be indicated; if the mouth is open and there is much saliva, Ammonium carbonicum may be helpful. If there is a cough, see Heart cough, above. Herbal Dandelion leaf can help to reduce fluid accumulation.

WESTIE LUNG DISEASE (IDIOPATHIC PULMONARY FIBROSIS)

Some cases have been successfully treated using homeopathy. Seek advice from an experienced holistic veterinarian.

See also *Supplements (22–23)*

URINARY PROBLEMS

Urinary problems usually result in changes in urine or urination. There may be blood or pus in the urine, or straining or pain on urination; urination may be more frequent or there may be involuntary urination. An inability to urinate is a medical or surgical emergency.

◀ *Frequent urination by a male dog can represent normal territory-marking behaviour.*

BLADDER STONES (UROLITHIASIS)

This is a metabolic condition, resulting from dietary imbalances. A natural and species-suitable diet is likely to be the best preventive. However, should it occur, homeopathic Berberis may help to prevent the formation of new stones and may even assist in dissolving those present. Surgery is sometimes necessary, in the first instance, especially in male dogs. Herbal Eupatorium purpureum may also be helpful. Good water intake is essential, to maintain urine flow. In Dalmatians there is a specific inherited condition that can give rise to stones, in which nitrogen waste cannot be processed correctly. Homeopathic Benzoic acid or Uric acid can often help these dogs.

BLADDER TUMOUR

Homeopathy has helped cases of malignant bladder tumour, as shown by biopsy and serial ultrasound scans. Bladder polyps are benign growths and can sometimes be cleared using homeopathic Thuja. If there is bleeding from polyps, homeopathic Nitric acid may help. See also pages 88–89.

BLOOD IN URINE (HAEMATURIA)

Blood can appear in the urine from anywhere in the urinary tract, including the kidney, bladder, ureter, urethra or prostate. If the blood is from kidney degeneration, homeopathic Phosphorus is well indicated. If clotted blood is produced, homeopathic Ipecacuanha may help. If the blood is from cystitis, see below. If the blood is from a tumour or polyp, see above. If the blood is from the prostate gland, see pages 80–81.

CYSTITIS

Cystitis is inflammation of the bladder. This usually occurs in conjunction with bacterial infection. There may be pus. If there is pain and straining at urination, homeopathic Cantharis may be helpful. Cleavers, Dandelion root, Parsley, Barley water or Cranberry juice are useful herbal preparations to help bladder health. Good water intake is essential to maintain urine flow.

INCONTINENCE (ENURESIS)

Involuntary urination can occur in spayed bitches as a relatively common delayed side-effect of the surgery. In some cases the enuresis can recur at six-monthly intervals, according to the female cycle that existed prior to spaying. Homeopathic Causticum can be helpful. In some cases either a homeopathic preparation of oestrogen or even a low material dose (prescribed by a vet) is necessary, for restoration of good control.

KIDNEY FAILURE

Kidney tissue, once lost, cannot be repaired, so this condition is life-threatening. It can be quite advanced before signs (thirst, characteristic smell, loss of weight) show. The dog can only survive if there is sufficient viable kidney tissue left at the onset of treatment. Homeopathic medicines such as Carbamide, Kali chloricum, Mercurius solubilis or Phosphorus, prescribed according to the individual patient, have been of great benefit. Fluid therapy may be essential at the outset of treatment.

POLYURIA

If a dog is urinating more copiously than normal (and drinking more), this can be a sign of serious disease, such as diabetes mellitus, diabetes insipidus or Cushing's Syndrome. Consult a vet for a diagnosis, and see also pages 86–87.

▶ *Polyuria and Polydypsia (PUPD) can indicate kidney or other serious disease. Avoid using plastic water or feeding bowls.*

REPRODUCTIVE-SYSTEM PROBLEMS

Reproductive problems can result in changes in behaviour, failure to breed, abnormal discharges, irregular female cycles, changes in the mammary glands and, in the case of pyometra (an emergency condition), increased thirst.

FALSE PREGNANCY

A bitch will always experience the hormonal changes consistent with pregnancy for the full nine weeks after each 'season' or 'heat'. However, in most cases this hardly shows except as slight changes in mood, demeanour and behaviour. When it becomes really obvious, it can be very distressing and may even give rise to full milk production, which persists as a result of the bitch sucking herself. A good, natural, fresh diet is likely to reduce these problems in the longer term. Homeopathic treatment can provide relief, relying on the constitutional approach for each individual. Homeopathic medicines such as Cyclamen, Lachesis, Lilium tigrinum, Pulsatilla or Sepia have been used. If there is milk production, do not massage the glands. Providing a 'body stocking' made from a pair of tights may help, by preventing the bitch from sucking and licking. Homeopathic Urtica (low-potency) or herbal Dandelion leaf may help to reduce secretion.

HYPERSEXUALITY (FEMALE)

This is rarely a problem, but homeopathic Lachesis, Lilium tigrinum, Murex or Pulsatilla may help, prescribed according to individual characteristics.

HYPERSEXUALITY (MALE)

This is not uncommon. Homeopathic Gelsemium, Lycopodium, Picric acid or Tarentula hispana may help, as may the relaxing essential oil Lavender, the Bach Flower remedies Holly and Chicory or the herbal calmative Valerian. If more drastic treatment is required, consult your vet.

INFERTILITY AND DECREASED SEXUAL BEHAVIOUR

Diet is obviously very important in both sexes. Homeopathic medication can help some cases of infertility, using remedies such as Agnus castus, Lachesis, Natrum muriaticum, Pulsatilla or Sepia, prescribed according to individual characteristics.

LACTATION

If milk is insufficient, homeopathic Pulsatilla and high-potency Urtica or herbal Milk Thistle and Goat's rue can often remedy the situation. Diet is very important. If milk persists after weaning, low-potency homeopathic Urtica or herbal Dandelion leaf can help to reduce it.

MAMMARY TUMOUR

In general, surgical removal of malignant tumours can lead to rapid growth of lung

▲ *Normally attentive bitches and dogs can be distracted and wayward under the influence of hormonal changes.*

secondaries and death, within a few months. Benign growths rarely need removal unless they become too large for comfort (they tend to grow with each reproductive cycle, so spaying is often recommended by vets). In rare cases, benign tumours can become malignant. See also pages 88–89.

MASTITIS

Inflammation of the mammary glands can occur at any time, but usually either during or after the nursing period or during the lactation phase of false pregnancy. Belladonna (hot, red, sore), Bryonia (swollen and painful on movement) and Lachesis (if purpling) have been found to be very useful homeopathic medicines.

PREGNANCY

Homeopathic Caulophyllum, given about six times during the last three weeks of pregnancy, is likely to help the subsequent birth process. Homeopathic Calcarea phosphorica, given similarly and once daily for the first week after delivery, is likely to prevent eclampsia.

PROSTATE-GLAND ENLARGEMENT

The prostate gland can become swollen and thereby lead to difficulty with defecation. This can also be a painful condition. Homeopathic or herbal Sabal serrulata can reduce the prostate. If it is painful, homeopathic Apis, Belladonna and Pulsatilla may help. Castration is often recommended by vets.

PYOMETRA

Pyometra (inflammation of the uterus) is usually an emergency condition and veterinary help should be sought promptly. With the guidance of a holistic vet, homeopathic Aletris, Caulophyllum, Lilium tigrinum or Sepia may help to prevent surgery. However, if relief is not rapid, surgery becomes necessary to prevent serious and life-threatening complications.

WHELPING

There are many natural medicines (herbal and homeopathic) that can help birth complications, but this is the realm of the holistic vet. Homeopathic medicines such as Calcarea phosphorica, Caulophyllum, Cuprum aceticum, Gossypium, Pulsatilla or Secale may be indicated.

See also *Spaying (44–45), Castration (46–47)*

BEHAVIOURAL ISSUES

Behaviour is but a mirror of a dog's emotions and, when observing behaviour, the best we can do is guess at the mental process behind it. If your dog exhibits any signs of the problem behaviours mentioned below, you should discuss it with your vet or a dog behaviourist. The list offers some effective natural remedies that may help with the behavioural issue in question.

A GUIDE TO NATURAL REMEDIES FOR BEHAVIOURAL ISSUES	
Issue	**Therapies and remedies**
Aggression	**Homeopathy:** Belladonna, Lycopodium, Nitric acid, Nux vomica. **Aromatherapy:** Lavender, Chamomile. **Bach Flower:** Heather, Vervain, Vine, Willow. **Herbs:** Hops, Skullcap, Valerian.
Anxiety	See *Timidity.*
Barking	See *Aggression, Separation anxiety.*
Bereavement and grief	**Homeopathy:** Cyclamen, Ignatia, Natrum muriaticum, Phosphoric acid.
Coprophagy (faeces-eating)	**Homeopathy:** Hyoscyamus, Sulphur, Veratrum album.
Destructiveness	**Homeopathy:** Belladonna, Chamomilla, Hyoscyamus, Stramonium. **Aromatherapy:** Lavender, Chamomile.
Excitability	**Homeopathy:** Belladonna, Hyoscyamus, Ignatia, Magnesium phosphoricum, Pulsatilla, Stramonium. **Aromatherapy:** Lavender, Chamomile. **Herbs:** Hops, Skullcap, Valerian.
Fears	**Homeopathy:** *Noise fear:* Nitric acid, Nux vomica, Phosphorus; *Anticipatory fear:* Argentum nitricum, Gelsemium, Lycopodium, Silicea; *Motion:* Bryonia; *Touch:* Arnica, Chamomilla, Lachesis, Nux vomica; *Thunder:* Borax, Gelsemium, Natrum carbonicum, Phosphorus, Rhododendron, Theridion. **Aromatherapy:** Lavender, Lemon balm. Bach Flower Aspen, Mimulus, Rock rose. **Herbs:** Hops, Skullcap, Valerian.
Hyperactivity	**Homeopathy:** Arsenicum, Coffea, Ignatia. **Aromatherapy:** Lavender, Chamomile. **Herbs:** Hops, Skullcap, Valerian.

◀ *Fearful dogs may require homeopathic, herbal or Bach Flower treatment, to help improve quality of life.*

Issue	Therapies and remedies
Hypersexuality	See pages 80–81.
Indifference	**Homeopathy:** Natrum muriaticum, Platina, Sepia. **Bach Flower:** Clematis, Wild Rose.
Jealousy/ possessiveness	**Homeopathy:** Apis, Hyoscyamus, Lachesis. **Bach Flower:** Holly.
Panic	**Homeopathy:** Aconitum, Gelsemium, Stramonium. **Aromatherapy:** Lavender. **Bach Flower:** Rock Rose. **Herbs:** Hops, Skullcap, Valerian.
Rage	**Homeopathy:** Belladonna, Hyoscyamus, Lyssin, Stramonium. **Aromatherapy:** Lavender. **Bach Flower:** Holly, Vine. **Herbs:** Hops, Skullcap, Valerian.
Resentment	**Homeopathy:** Natrum muriaticum, Staphisagria. **Bach Flower:** Willow.
Restlessness	**Homeopathy:** Arsenicum, Coffea, Ignatia, Rhus toxicodendron, Stramonium. **Aromatherapy:** Lavender, Chamomile. **Bach Flower:** Impatiens, Vervain. **Herbs:** Hops.
Separation anxiety	**Homeopathy:** Lycopodium, Phosphorus, Pulsatilla **Aromatherapy:** Lemon balm. **Bach Flower:** Aspen, Cerato, Larch, Mimulus.
Timidity/want of confidence	**Homeopathy:** Ignatia, Natrum muriaticum, Pulsatilla, Silicea. **Aromatherapy:** Lavender, Lemon balm. **Bach Flower:** Larch, Mimulus, Vervain.
Wandering/ roaming	**Homeopathy:** Bryonia, Sulphur, Tub. bov.

ALLERGY AND ATOPY

Allergy and atopy are common canine conditions and possibly account for most of the visits to a vet. The approach to the problem taken by holistic and natural medicine is not the same as that commonly recommended by more conventional medicine.

The term 'allergy' is derived from the ancient Greek, meaning 'other function' (referring to the incorrect functioning of the immune system). It is therefore one of the few modern disease names that accurately and meaningfully describes the fundamental problem, rather than simply describing the signs and symptoms. Atopy is a term used to describe a generalized allergic or hyperallergic condition.

COMBATING ALLERGENS

In allergy or atopy, the immune system responds inappropriately to 'allergens', which are external materials that 'trigger' the reaction. These may be pollens, house dust, gluten, flea saliva, moulds, grass proteins, plants, food components, household reagents, agrochemicals or certain other chemicals. A common sign is itchiness of the skin, but diarrhoea, vomiting or respiratory problems are also seen quite regularly.

Removing or reducing these allergens from the environment or diet of the dog may temporarily reduce the symptoms, but will not effect a 'cure'. It has to be a lifelong endeavour – a daunting prospect. Furthermore, if the allergen is an environmental factor that cannot be removed (such as house dust), this is an impractical option. Even if it were possible to separate the dog completely from the supposed allergen, the immune disturbance might enable an allergy to other environmental items to develop. So the underlying immune malfunction must be corrected before the dog can be free of allergies.

Any powerful immune challenges that are capable of perverting immune

◀ *Allergy or atopy will often produce an itchy skin.*

REMEMBER...

Allergy and atopy result from immune disturbance. If this disturbance can be corrected, the allergy will fade and the signs and symptoms will subside.

Immune disturbance can be caused by chemicals, viruses or vaccination.

▲ *Many household chemicals can be a problem for an allergic dog.*

function and balance, such as viruses or immunization (vaccine), are likely candidates. Certainly, revaccination of an allergic dog is not recommended.

Desensitizing vaccines may be offered conventionally, which are based on the result of specific allergy tests. This expensive technique seems to be aimed at tiring, exhausting or wearing out the immune system into non-reactivity, but only appears to help in a few cases.

THE NATURAL APPROACH

Rather than trying to suppress the allergic reaction with anti-inflammatory drugs such as steroids (cortisone) or with anti-histamines, for life, which can result in a deepening of the underlying disease, natural medicine – particularly the homeopathic constitutional approach – can apparently bring about a lasting rebalancing of immune function and a cessation of allergy.

It is wise to try to reduce the environmental challenge in the meantime, but – whereas in conventional medicine this is a permanent requirement – in natural medicine a normal tolerance for the allergen may be re-established.

Diet is an essential component of holistic treatment, in that the immune system should heal much faster and more effectively if given a natural, healthy, unprocessed and species-suitable diet. Furthermore, some allergies to dietary chemicals occur, for which the animal has no need, but which may be harmful anyway.

Homeopathic medicines that are common constitutional prescriptions for allergic dogs include Bacillinum, Calcarea carbonica, Graphites, Kali sulphuricum, Lycopodium, Phosphorus, Psorinum, Pulsatilla, Sepia, Sulphur, Thuja and Tub. bov. These are prescribed by the holistic vet on the basis of constitutional indicators. The bowel nosode Morgan Bach is also commonly indicated.

See also *Vaccination and Alternatives (34–37), Gastrointestinal Problems (66–67), Skin Problems (68–69), Chest and Heart Problems (76–77), Autoimmune and Endocrine Disorders (86–87)*

AUTOIMMUNE AND ENDOCRINE DISORDERS

The endocrine system is a closely integrated system of glands (including the thyroid, adrenal, pituitary and ovaries or testes) that secrete hormones into the bloodstream to regulate the body's responses to external influence. The immune system is a complex of structures and mechanisms that protect the body from infectious disease.

▲ *Many autoimmune disorders can be labelled 'hereditary'. This may be erroneous.*

Autoimmune disorders are those in which the dog's immune system has misrecognized the body's own tissues as 'enemy' and causes damage to those tissues. Many differently named conditions come under this umbrella heading. Most of the endocrine disorders that dogs suffer may also be of an autoimmune nature. While a poor prognosis is usually offered for these conditions, many cases have recovered well on homeopathic treatment.

While the actual causes of autoimmune disorders are not established for certain, it is thought that factors that can impinge on immune function are likely to be involved. Virus challenge, vaccination and dietary anomalies may give rise to an autoimmune condition. As vaccines often contain residues of animal tissues, it is not difficult to see how they may provoke a wayward immune response (recognition malfunction) of this nature in some dogs.

CANINE AUTOIMMUNE DISORDERS

Named autoimmune syndromes in dogs include: Autoimmune Haemolytic Anaemia (AIHA), dry eye (KCS), Vogt-Koyanagi-Harada Syndrome (VKH), pemphigus, lupus, Sudden Acquired Retinal Degeneration (SARD, SARDS), eosinophilic myositis, Chronic Degenerative Radiculomyelopathy (CDRM/DM), thyroid disease and heart-valve disease (of Cavaliers).

Different breeds show different tendencies when pushed into an autoimmune state. Westie lung disease (idiopathic pulmonary fibrosis), Addison's Disease, Cushing's Syndrome and diabetes mellitus may have autoimmune aetiology. It is not impossible that cardiomyopathy may be included in this grouping, when it is better understood.

In conventional practice, immunosuppression is usually considered, by means of high doses and prolonged courses of steroid (cortisone). This is not without its dangers. However, many cases appear to respond very positively to homeopathic treatment with holistic support, in the shape of a natural, fresh, organic diet. At least recovery is coincidental with homeopathic input. These successful cases appear to show that homeopathy can provide a realistic alternative to suppressive drug therapy. The approach used is the homeopathic constitutional approach, tailored to the individual patient rather than to the name of the disease. This is best performed by a holistic vet. Many cases on record seem to have enjoyed an apparent cure, despite the poor prognosis that is usually offered.

CANINE ENDOCRINE DISORDERS

Endocrine disorders that commonly occur in dogs (apart from those of the ovaries or testes, see pages 80–81) are hypothyroidosis, hyperthyroidosis, Cushing's Syndrome, Addison's Disease, diabetes mellitus and diabetes insipidus. In all of these there are clues that there may be an autoimmune component; as more research unfolds, this suspicion may in time be confirmed. There are homeopathic, dietary and herbal means of treating these disorders, with reasonable hopes of positive results – the patient should be under the care of a holistic vet.

See also *Vaccination and Alternatives (34–37), Reproductive-system Problems (80–81)*

CANCER: AN APPRAISAL

Cancer is a common disease, appearing to become more prevalent as the years go by. No particular reason for this is agreed upon, but diet, vaccination, chemicals and environmental degradation are possible players.

It is a disease in which certain cells grow and multiply in a way that is outside the body's normal hierarchical control. This can either result in growths (tumours), which are capable of spreading (metastasis), causing damage or even proving fatal in malignant types; or it can affect the blood or the lymphatic system. Cancer is very prevalent in dogs, with Flat-coated Retrievers appearing to have a particular susceptibility. In the USA one in three dogs can be expected to develop cancer, and it will be fatal in one in four. Cancer kills more dogs over two years old than any other disease.

Surgical excision, steroids or various combinations remain the usual options offered conventionally. If the tumour is malignant and invasive, surgical removal may lead to 'release' and rapid development of secondary growths that may already have started in a very small way. Particular examples are osteosarcoma and mammary carcinoma, in which secondary development (mostly in the lungs) is a frequent sequel to surgery. Not all types of cancer are considered amenable to chemotherapy. Interestingly, chemotherapy is not usually

REMEMBER...

Cancer is a serious disease, with one in four dogs expected to die of it. There is no consensus on definitive causation or on treatment rationales. However, an encouraging number of cases have recovered while using holistic and natural medicine.

▶ *A holistic veterinary consultation with holistic management is an excellent way to maintain health.*

taken as poorly by canine patients as it is by humans; nonetheless, there are generally deleterious side-effects, and some sensitive patients can be made very miserable indeed.

THE HOLISTIC APPROACH

In holistic medicine it is believed that if the body has the power and the will to fight the cancer, if its ability is appropriately stimulated (for example, by homeopathy) and if all obstacles to recovery are removed, then a positive outcome will ensue. This is borne out in real-life experience by a number of cases. While by no means all cases end well, recoveries have been recorded in most types of cancer. These include fibrosarcoma, osteosarcoma, mast-cell tumour, histiocytoma, lymphoma, mammary carcinoma, liver cancer and spindle-cell tumour. It even appears that – contrary to popularly held belief – homeopathy can exert a benefit in the face of chemotherapy or steroid treatment.

This means that you don't have to adopt a strictly either/or approach. A change to a healthy, natural, fresh, organic and species-suitable diet is an essential component of the holistic approach. Treatment by a holistic vet is recommended.

Some causative factors that appear to have been identified from past cases are injury and trauma (whether mental or physical), chemical exposure, so-called 'electro-magnetic smog', incorrect diet, viruses and vaccination. These factors may act singly or together. In this connection, it is interesting and concerning to note that many living vaccines are cultured on laboratory cell cultures that intentionally contain cancer DNA. In cats, there is a known risk of cancer developing at the site of injection, possibly for this very reason.

▲ *Chemotherapy is less traumatic for dogs but brings extension of life only in selected cases.*

In cancer, a positive outcome is not an impossible target. There are grounds for reasonable hope. Recovery has occurred in an encouraging number of cases. However, it is clear that cancer is a particularly vicious and powerful enemy, and a great deal of work needs to be done on methods of prevention. The prognosis offered at the outset of treatment of any single case has to be very guarded.

See also *Supplements (22–23)*

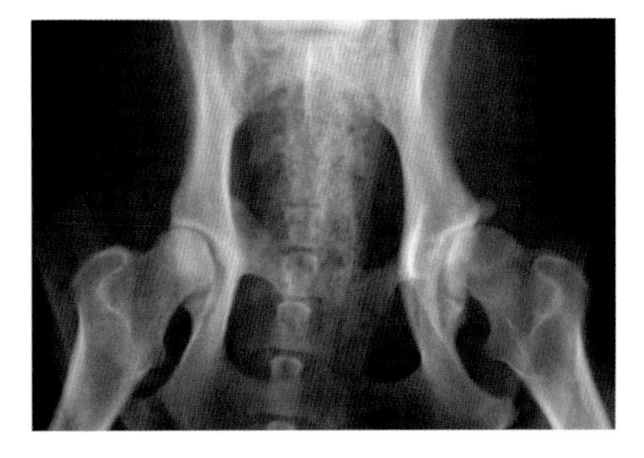

▲ *Arthritis is a common ailment of older dogs.*

cold, bright weather and better for warm and damp conditions; it also has strong indications in the weakness that some older dogs suffer, and may help with Vestibular Syndrome, age-related deafness, incontinence or laryngeal paralysis. Baryta carbonica can suit the old dog who becomes confused and 'absent'. Ambra grisea is well suited to general ageing and weariness. If there are heart problems, consider Crataegus, in 'mother tincture' form, as a first line of treatment; and Vitamin E may help. If there is a 'heart cough', Apis mellifica or Spongia tosta may have a part to play. In kidney disease, Mercurius solubilis, Phosphorus or Kali chloricum may be needed. If the hind legs are giving way, with no obvious pathology and no X-ray signs, Conium can prove useful.
- In herbal medicine, Hawthorn berry can support the heart and Dandelion leaf is a great stand-by as a herbal diuretic. Willow-bark, Nettles, Devil's claw, Meadowsweet and Cleavers can help with arthritis and limb problems.
- Regular acupuncture can provide immense benefit for the aged dog, acting as a general balancer or helping with medical problems such as back pain, arthritis or paralysis.
- Regular chiropractic checks will help to keep the musculoskeletal system aligned and prevent back pain.

▶ *Nettle is one of the herbal medicines that can be used in cases of arthritis.*

The above list of treatments is far from exhaustive, and it is advisable to seek the advice of a holistic vet for help with age-related problems. Should cancer develop, it is not impossible for the dog to recover, given the help of homeopathy, herbs and a good diet (see pages 88–89).

EUTHANASIA

Towards the end of a dog's life there is always the difficult question of euthanasia. However, just because it is possible to 'put

him out of his misery' does not always make it right. It is not unusual for a dog that has been raised, fed and managed according to holistic and natural principles to find his own way out of life, with dignity and peace. Death does not have to involve suffering; after all, it is a natural and inevitable part of every life.

One course of action to consider, if the circumstances are correct, is giving homeopathic medications that can help a dog make up his mind whether to go or stay (ask your holistic vet for advice). Giving these does not constitute euthanasia, but can help the dog either pick up in health and energy or decide to leave this world, in which case a peaceful passing often seems possible.

Decisions on whether or not to perform euthanasia are very difficult. In general, if an animal is suffering significant loss of quality of life and dignity, and if there is no hope for recovery and he appears unable to find his own way out calmly, then intervention becomes necessary.

When a decision is made, vets will often make a house-call, to enable a dog to die in his own home among friends. Sedatives may be given beforehand, if this seems appropriate. A lethal injection may then be given, usually into the bloodstream, via a vein.

In all cases where euthanasia is being considered, you are likely to find your vet sympathetic and willing to talk through the situation. Neither vet nor owner will wish to take a life incorrectly, so joint discussions are a support to both at such difficult times.

▲ *Older dogs can still enjoy walks and activity.*

See also *Natural Weight Control (26–27), Common Ailments (64–81), Cancer (88–89)*

INDEX

ACKNOWLEDGEMENTS

AUTHOR'S ACKNOWLEDGEMENTS

In preparing this book, I am very conscious of the magnificent support that I have received from my family and my publishers, without which I could not have completed the work.

I wish also to acknowledge the inestimable contribution made by my canine patients and their human families, who have helped me to learn to use natural therapies in veterinary practice.

RESOURCES

Academy of Veterinary Homeopathy (AVH): www.theavh.org

American Holistic Veterinary Medical Association (AHVMA): www.ahvma.org

International Association for Veterinary Homeopathy: www.iavh.org

International Veterinary Acupuncture Society: www.ivas.org

Non-commercial natural feeding advice: www.naturalfeeding.co.uk

PUBLISHER'S ACKNOWLEDGEMENTS

Executive Editor: Trevor Davies
Managing Editor: Clare Churly
Editor: Ruth Wiseall
Executive Art Editor: Penny Stock
Designer: Sally Bond
Picture Library Manager: Jennifer Veall
Picture Researcher: Emma O'Neill
Senior Production Controller: Amanda Mackie

PICTURE ACKNOWLEDGEMENTS

Alamy/blickwinkel/Schmidbauer 81; /Katrina Brown 85; /Kris Butler 53; /John Eccles 61; /Isobel Flynn 18; /Martin Fowler 71; /Juniors Bildarchiv 6, 43; /Adrian Sherratt 88; /Tierfotoagentur/B. Schwob 84; /Westend61 GmbH/Creativ Studio Heinemann 52. **Ardea**/John Daniels 77; /Jean Michel Labat 66. **Barcroft Media** 51. **Corbis**/Dale C. Spartas 91; /Inspirestock 27; /Tetra Images 62. **Dorling Kindersley**/Tim Ridley 58. **T. J. Dunn, DVM** 92 top. **FLPA**/Minden Pictures/Mark Raycroft 1. **Fotolia**/ChinKS 44; /Frog 974 33; /Angelika Möthrath 37; /Dmitrijs Novikovs 41. **Getty Images**/Gerard Brown 86; /Dorling Kindersley 70; /Yellow Dog Productions Inc. 47. **Nature Picture Library**/Jane Burton 36, 60. **Octopus Publishing Group**/Stephen Conroy 23, 25, 31; /Janine Hosegood 29; /Sandra Lane 30; /Russell Sadur 9, 14, 15 left, 15 right, 16, 17, 19, 54, 67, 74, 83; /Ian Wallace 22. **Photodisc 8. Photolibrary Group**/Banana Stock 21; /Big Cheese 7; /BSIP Medical/Hubert Hubert 39; /Fotosearch 4; /Garden Picture Library/Howard Rice 73; /Garden Picture Library/Janet Seaton 48; /Garden Picture Library/ Jo Whitworth 92 bottom; /imagebroker.net/Picani Picani 13; /Juniors Bildarchiv 10, 24, 26, 34; /Olive Images 78; /Oxford Scientific (OSF)/Nick Ridley 68; /Photodisc/White 12; /Phototake Science/Dennis Kunkel 40; /Pixtal Images 11; /Purestock 79; /SuperStock/Jerry Shulman 2. **RSPCA Photolibrary**/E. A. Janes 93; /Becky Murray 63. **Science Photo Library**/Eye of Science 42; /Marilyn Schaller 38. **Tom Thompson** 89. **Tips Images**/ Bios 50.